The Ultimate MAZDA Performance Guide

(2004+ Models)

By Derrick Ambrose

Why Modify?

- Performance
- Handling
- Safety
- Fuel Economy
- Extend Your Mazda's Life
- Upgrade Worn Out Parts

TABLE OF CONTENTS

INTRODUCTION

Unless you are a billionaire, building the car of your dreams is almost impossible. Starting with a Mazda and crafting it to a car that can exceed your expectations, however, is a much less daunting task. This book was written to help you succeed in learning the ways to elevate your Mazda to those levels. With the aftermarket parts market growing every year, trying to determine which route to take while building the car you want can be intimidating. In a burgeoning sea of options and opinions about which modifications (mods) are worthwhile and what their best order would be, I hope to empower you to take advantage of the many opportunities available to transform your Mazda and, at the same time, provide you with the peace of mind knowing you are doing it the right way.

Only a few dozen products are mentioned in this book. They have been culled from my decades of experience in modifying Mazdas and are considered by me to be the best of the best. These modifications offer the best bang for the buck and have been proven to deliver time and time again, by countless tuners whom I have worked with around the world. This book is tailored for you, the beginner, to get your foot in the door without spending thousands of dollars and dealing with the usual headaches.

What do I consider a beginner? I'd say if you've purchased and installed five to ten of the mods in this book, then you probably have enough experience to be considered a resource in the Mazda community. If you have installed half of the products mentioned here, your skills are probably beyond the scope of this book. By using this book, the novice can save a lot of time spent in face-to-face inquiries with "experts" or getting questions answered on many internet forums, which can be sometimes difficult at best. People can be very rude to the inexperienced user, who is merely trying to ask a simple question, such as, "What is an SRI?" The response to that question and many others will be answered here, giving you the education and confidence to enter a forum and contribute from day one without sounding like a "noob".

Enjoy,

Derrick Ambrose

CHAPTER ONE

WHY MODIFY?

"Why would someone ever do that to their car?" At some point in your life, you've likely been driving down the street and seen an obnoxiously loud Honda pass you with their stereo thumping and their car sitting so low its unclear how they can even drive it. Maybe you've seen cars go in a driveway at an angle so they can make it into a parking lot with their exhaust or bumper still intact; or the 4x4 truck so tall, you would need a ladder to get in. On the other side of the spectrum, there, probably, has been a time or two when you have seen a car that has all the right types of modifications to make it sleek, sexy and powerful and you thought to yourself, "Nice ride! That is exactly how I want my car to look."

The point is, not all modifications are created equal. In the pages that follow, I will go into deeper detail about some smart modifications you can make to your vehicle that will give you an optimal balance of style and performance to make your Mazda uniquely yours. But before I do, I will touch on the overarching question. Why modify?

Like your clothing or your shoes, your Mazda says a lot about who you are. It's an extension of your personality. For many, mods are an outlet to express oneself and add a personal touch. For others mods are about saving money by improving gas mileage or making their car safer to drive. For others still it is all about improving power or handling.

There are many ways you can choose to modify your Mazda, and the importance of knowing what you hope to achieve with your modifications before you start making them, cannot be understated. Do you want to win car shows, get on the cover of a car magazine, have the loudest stereo, make 1,000 horsepower (hp) or improve handling? Or, for more practical reasons, do you simply want to improve fuel efficiency or safety? It is important to determine if you want to modify and, if so, why you want to modify, before you begin to consider what parts make the most sense to purchase. You should not add parts simply because your neighbor or friend has them on their car. That type of approach, most likely, will leave you disappointed and wishing you had read this book.

Before I get started diving into the details that will help you down the path to perfecting your Mazda, I will discuss some of the many reasons you should consider modifying your car. Let's dig in.

Reason #1a: Power

There are countless options on the market that can aid in increasing the power of your Mazda. Some of the more common modifications include intake systems, exhaust systems, and engine tunes. But, why should you increase engine performance? Increasing engine performance on your Mazda increases its efficiency. This results in better vehicle responsiveness, such as increased torque and horsepower. You can get even more performance from items that do not directly generate extra horsepower. Lighter engine components like flywheels and pulleys can make it easier for your engine to rev up.

Why Modify

- **More horsepower.** This means quicker acceleration and increased efficiency from your engine. Have a full load of people in your Mazda? Need more load capacity for towing? An extra 10-15 horsepower is something that will make a noticeable difference in day-to-day driving. I guarantee you will be able to feel it when you take off from a stoplight.

- **Improved responsiveness** (throttle response) results in less pedal pressure needed to get the same response from your engine.

- **Safety is improved by adding power.** More power means it's easier to merge and maintain speed going up steep hills.

- **A more pronounced engine sound.** The audible queues from your engine make it easier to match revs for a smooth clutch engagement on manual transmission-equipped cars. Plus, who doesn't like to hear a bellowing engine under full throttle? It's music to the ears.

Reason #1b: Handling

Modifying the suspension on your Mazda also has many benefits. For the most part, when people think about suspension modifications, they think about lowering springs or coilovers. Adding these parts will enhance the look of your Mazda, but that is not the only reason to consider suspension upgrades.

Why Modify

- **Give your car a more aggressive, sportier stance.** Adding lowering springs or coilovers to your Mazda not only will improve handling, but also, it will give your Mazda that lean, aggressive and sleek look that screams sexy.

- **Decrease Body Roll.** No, I do not mean the rolls above your jeans; the gym can handle that. What I mean by "body roll" is the load transfer from one side of your vehicle to the other, when turning. With decreased body roll, comes the ability to turn quicker and have more confidence while you do it.

- **Get nice and stiff!** Performance shocks will decrease suspension compression and rebound. This will give you a stiffer ride. The stiffer your suspension set-up, the more you will feel those speed bumps in the grocery store parking lot, but the more you will feel the road, allowing you to react quicker. There are many different types of suspension set-ups...more on that later.

- **Keep your chassis rigid. Stay smooth.** Your chassis is constantly twisting and contorting while you're cruising down the road, especially during cornering. By adding extra bracing to your chassis, you will maintain your suspension's optimal positioning. This translates to better feedback under cornering, informing you, the driver, of what is happening with your Mazda.

- **Safety** is a fantastic benefit of handling upgrades. We've all had to dodge debris in the road, an animal crossing the road, or the F-150 that didn't catch you in their blind spot. Handling upgrades allow you to maintain better control of your Mazda during these evasive maneuvers.

Body roll is a reference to the load transfer of the vehicle toward the outside of a turn. Think of riding a city bus and the lean you feel when it corners.

1.2 Styling

Styling components, often, are the easiest modifications to spot on a vehicle. Without a certain level of restraint, they can become tacky and, actually, make your car look cheap. With the right balance, your vehicle can become an extension of yourself; it can look awesome and speak volumes about your lifestyle.

If your car could talk, what would it say about you?

> *Minivan:* "I've given up my (own) life. It's all about the family now: cargo over charisma. Maybe, one day, I can get a Miata.

> *Crossover:* "Yep, I have a family, but I'm not quite ready to totally sacrifice style. Check out these yoga pants."

Sedan: "It's called a Win-Win, or compromise. She didn't get the minivan, and I didn't get the sports car."

Hatch/Compact: "I'm young and looking for a good time. Did I mention the back opens up wide, so I can put my gear in there?"

Sports Car: "I don't have to worry about practicality. This is form over function, well, at least the kind of function that gets you from point A to point B."

The type of car you drive and the way you style it says a lot about who you are. Right or wrong, people make assumptions about you based upon the visual cues you provide. It is human nature.

By adding custom wheels, tint, LED's, body kits, shift knobs, and other exterior and interior parts, you can customize your Mazda to match your lifestyle. It's the difference between a custom-tailored suit, made specifically for you, and one you get off the rack at Macy's.

As a special bonus, some styling mods, done right, can give a performance boost, albeit at 100+ MPH. So, tell people you did it for the aerodynamics.

1.3 Reason #3: Safety

Safety is not something people usually associate with modifying, but there are plenty of performance upgrades that can make your Mazda safer, such as a roll bar for a Miata.

- Upgraded braking components provides more responsiveness; and many brake pads have higher temperature capabilities to survive constant high speed braking. This is great, if; for example, the person in front of you slams on their brakes on the freeway, or when you are braking on hills driving through a pass, or carrying heavier loads, such as towing a trailer. Better brake pad compounds give you the ability to stop faster in emergency situations.

- Better suspension components can make the difference between a traffic collision and a close call. Think about the difference you would expect to experience if you were trying to dodge a mattress on the freeway in a Miata versus an MPV. With upgraded suspension components, dodging that Serta will be much easier.

- Have you ever wished for a bit more power when you were merging into traffic on the freeway? An increase of 15-20 horsepower can get you up to freeway speed 30 feet sooner or allow you to pass in a shorter distance.

1.4 Reason #4: Fuel Economy

Think about drinking a Slurpee on a hot summer day, except you grab one of those small coffee straws and try to drink. No matter how hard you suck, you will have a hard time getting any of the Slurpee out of the cup. Now, imagine you grab a large straw. Suddenly, it is much easier to gulp down that delicious drink. The same concept applies to your intake or exhaust. The freer flowing the system, the less hard your car will have to work to get the same result. By removing restrictions that slow the flow of air and limit exhaust gases from escaping easily, your engine is able to perform more efficiently.

TEST RESULTS

My testing of the SkyActiv CX-5 showed an average of 34.0 MPG on the highway with the factory intake system. After installing a short ram intake on the car, the MPG improved to 38.5 MPG, on average. This represents a 13 percent increase in freeway fuel mileage with just an intake replacement. For you math majors, that means the intake can pay for itself in about 13,000 miles at current gas prices.

Of course, with improved power, can come a heavier foot which can offset any improvement in miles per gallon (MPG). If you are planning to make modifications for MPG improvement, be aware that more spirited driving will cause any improvement in fuel economy to disappear; so, before you put the pedal to the metal and race that Honda next to you off of the green light, you may want to think twice about why you added the mods in the first place.

There are a few gimmicks on the market that are designed to trick you into spending money on mods that don't live up to their promises. The sales pitch always seems to start the same. Install this "chip" or "tornado intake" and get an increase of 50% performance and fuel economy. Before investing money on any of these, do a little research. Wouldn't everyone do this if they were getting that type of performance improvement? The answer is, of course, that they don't work, and if it seems too good to be true, it probably is.

1.5 Reason #5: Vehicle Maintenance and Extending the Life of Your Vehicle

Your vehicle is one of the biggest investments you will make in your lifetime, and protecting that investment, by taking measures to extend the life of your vehicle, is a practical and smart reason to make modifications to your car.

- Replacing your filter with a performance filter is an inexpensive way to add life to your vehicle and improve gas mileage. A dirty filter affects fuel economy by making your engine work harder.

- Road salt, sludge, and pollution can lead to costly body work. By adding protective products such as skid trays and mud flaps, or replacing components such as exhausts with parts made from higher quality materials, you can help limit corrosion of your underbody and save thousands in rust-related repairs. This is particularly true in locations near the ocean and wherever roads are salted during the winter.

- Engine performance parts not only increase power output for your vehicle, but can also help to maintain your engine. Free flowing exhaust systems reduce backpressure and can aid in keeping the exhaust gas temperatures down which can help extend the life of your engine and its components.

- Customizing your car personalizes it. You'll love it more, which will increases the likelihood you'll hang onto it, rather than swapping it out when the next rodent on TV tells you it's time for a new car.

- High Performance Lubricant: Using high performance motor oil in your car improves performance and longevity in three key ways: It helps to keep the engine clean, minimizes friction, and combats wear. Normal mineral oils can cause dirt particles to stick together and form sludge on engine surfaces. This can prevent proper lubrication, which could lead to premature engine failure.

1.6 Reason #6: Upgrading Worn Out Parts

As your Mazda gets up in miles, parts simply wear out. Replacing your worn-out stock parts with upgraded performance parts will pay dividends in the long term. Good examples of this are the shocks and struts, which can wear out by 60,000 miles or less. Buying a set of new parts from Mazda can cost more than purchasing a set of performance aftermarket parts, and you'll get more value from the performance part.

Mazda engineers some amazing cars, but from time to time, you may find a component that is known to fail in a certain model or year vehicle. A perfect example of this is the side engine mount in the Mazda 3 and Mazdaspeed 3. Upgrade to a performance part that will last longer rather than a new stock part that, very well, may fail on you again.

The option to upgrade worn-out parts is available for many parts in your Mazda, from clutches to brakes, tires, wheels, lights, and more. Why go with the stock parts when you can get something better?

Conclusion

I've just outlined many of the reasons you may want to modify your Mazda. You don't have to make your muffler sound like an old lawn mower. Even the everyday driver, with a less conspicuous attitude has good reason to consider aftermarket modifications. Besides simply increasing power, there are many reasons to modify your car, and you may notice that a lot of the areas discussed in the previous pages will have positive effects in multiple areas. For example, you may get an intake to gain more gas mileage, and then, after installing it, you find that you've also acquired a few extra horses, enough to feel the difference.

Now that we've discussed why you should want to modify your Mazda, I'll present what you need to know before you begin modifying, so that you can pick the mods that best fit your needs.

CHAPTER TWO

BEFORE YOU BEGIN

It's a good idea to understand where you'd like to end up with your Mazda before you start spending money. Always begin with the end in mind. Do you just want to add an intake for improved fuel economy or would you rather go all out with dozens of mods? No matter how large or small your build, it's important to know where you're going before you start. In this section I will address some of the questions about what you need to know before you start your build.

Take a realistic look at what you plan to achieve, when you hope to achieve it by, and what type of budget it will likely take to get there. Once you understand the goals you have for your Mazda, this book can help you plan out your build.

You should also understand which mods "play" well with others. You should always try to add mods in a way that allows them to complement each other. Also consider your local laws. Some mods that may be fine in one state, may not be legal in another.

In the pages to follow I'll explain in more detail how you can achieve your goals in a comprehensive way that is not difficult. I will also answer some of the most common questions people ask when they are first getting started modifying their Mazda. I'll discuss common problems to watch out for and explain the meaning of a lot of the common acronyms and slang that you'll need to be familiar with as you get deeper into modifying your car.

2.1 What are your goals?

Do you want a show car, a family car, a track car, or something in-between? There are many different ways or "styles" you can choose to enhance your vehicle. You would probably get some strange looks and a few laughs if you had a lifted 4x4 truck with lambo doors and a giant exhaust. How about a lowered import car with mud tires and a winch on the front?

A great way to begin to understand the "mod" scene is to attend some local car shows and take a look at the different ways others have chosen to modify their cars. Spend some time looking at magazines and searching the internet. Looking at how others have modified their car will help you to determine what you like and what you don't.

Let's not forget to take into consideration how you use your vehicle. If you use it primarily as your daily commuter (daily driver or DD) then it probably isn't in your best interest to build it out as a 900hp drag car. By the same token, if you want to build a track car it may not be a good idea to put a 300 lb. sub-woofer box in the back and thousands of dollars in audio components. As cool as that stereo is, it isn't going to do you any good when you are on the track.

2.2 Budget Your Build:

Now that you have determined what you would like to achieve, it's time to take a realistic look at your budget. Make sure your goals and your budget are in line. If your goal is to make the cover of your favorite import car magazine and your budget is $1,500, you're probably going to end-up very disappointed. There are different ways you might want to go about determining your budget.

- **Make a list.** List all of the modifications you'd like to do and determine what the cost will be over the life of the build, then make a budget and timeline for your build. Unless you have a lot of expendable cash or do not want to do very much to your Mazda, it will take time to purchase all of the components.

It is best to list out the parts in order of what will get you the most bang for the buck first and start down the line until you have achieved your goals. I have laid this out for you in the next chapter.

- **Determine Your Max Budget Upfront.** Much like gambling, modding can get addictive! Decide what you want to spend and then prioritize your mods to fit within that budget.

- **Finance it.** Not enough cash in your paycheck to get that big upgrade you want? There are several options on the market like layaway and financing that can give you the flexibility needed to purchase a big item and not have to lay out the cash all at once.

Always remember to keep your factory parts stored somewhere handy as you replace them with aftermarket components. That way when you are ready to sell your car, you can take it back to stock if you need to.

2.3 The Best Platforms for Mods

Some Mazdas have lots of aftermarket parts available and some have virtually no aftermarket support, so it is important to take into consideration how much you would like to modify before you decide which Mazda to purchase. Knowing a little bit in advance about what is available in the aftermarket for the model you are looking to purchase can save you money later on, especially if your goals don't fit your car. To help you find the best car for your needs and wants we developed a rating system based on "modability".

RATING SYSTEM

Next to each car description you will see a rating from one to three. One represents a car with little to no aftermarket support. Two represents a car that has mods available but maybe not for everything you would like to do. A three would be a car that has lots of support for mods and lots of potential to make the car your own.

Mazda 2

This car is compact, has good gas mileage, is light weight, and fun to drive. There is even a race series called B SPEC that you can compete in. This lightweight car can be made into a fun car for very little money. The aftermarket is not heavily supported but it does exist, particularly for suspension parts. Intakes and exhausts are also available along with other small mods, but if big power is your goal then this is not the car for you.

Mazda 3

This car is still a compact but slightly larger and heavier than the Mazda2. It comes in a variety of models that respond very well to handling and power mods. They also get great highway mileage and the hatchbacks can hold loads of stuff in the back with the seats down. The newer SkyActiv engines are a great choice for gas mileage and respond to mods well. The larger motor 3's have great torque. There is a decent amount of aftermarket support for these cars but again if big power is your overall goal then you will likely want to look at the Mazda3's bigger brother the Mazdaspeed3.

Mazdaspeed 3

This car is only available in a hatchback version with a 6 speed manual transmission. It is a top of the line Mazda 3 with a much more powerful direct injected turbocharged engine. Everything you can do to a Mazda 3 can be done to a Mazdaspeed 3 and more. Aftermarket support for this model is readily available. It is common to see these cars at well over 300 wheel horsepower. This is the best Mazda on the market today to mod for power.

Mazda 6

This is a mid-sized car with great leg room, nice interior, and most of the creature comforts that you would expect to see in this size car. However when looking to the aftermarket, there is not a lot available. Like the Mazda 2, you can find intakes, exhausts and some suspension mods but if your goals include heavy modding then this would not be the best answer for you.

Mazdaspeed 6

This car uses the same engine as the Mazdaspeed 3, but is a larger, heavier, more luxurious car. This AWD car responds very well to modifications like the Mazdaspeed 3, but will get you there with a little more leg room. The AWD cuts down on the gas mileage but can be a nice benefit during the winter months or high rain seasons. Power mods are plentiful. Suspension and styling mods are harder to come by as it was only sold in 2006-2007.

CX-5

This is relatively new cross-over vehicle that is not heavily supported yet. There is a promise of the diesel engine making its way to the US which may help bolster the aftermarket potential. If you are looking for a cross-over that gets great gas mileage this car might be your answer.

CX-7

With the same engine as the Mazdaspeed3 and Mazdaspeed6 there are some aftermarket power options but that is about where it ends. If you are looking for a slightly mod friendly car that can haul the whole family, you are looking in the right place.

CX-9

This car is a large cross-over that has an almost nonexistent aftermarket but can hold 7 passengers.

RX-8

This sports car offers lots of aftermarket support and plenty of differences compared to your average car. It has great handling and a rotary engine, which makes this car very unique for a newer model Mazda. There are tons of upgrade potentials for this vehicle, but if you are looking for good gas mileage, look elsewhere.

Miata

This is the smaller lighter choice when it comes to a sports car. It is not your average two-seater and can be modified to the max. Every type of racing, styling and performance part you could think of, you can probably get. These cars are very popular in the autocross community or for people looking to get out of the house and blast the back roads on the weekend.

2.4 Top Ten Commonly Asked Questions About Modifying:

Over the many years I have spent on the front lines for a Mazda performance company, I've been asked a lot of very strange questions and some very practical ones as well. The following represent a list of questions that, it seems, just about everyone wants, should want, or needs to know before they begin their build.

1. What Should I Start With?

I've generally laid out chapter 3, "How to modify your Mazda", to answer this question. In short, if you are looking for the best bang-for-your-buck consider starting with the list below. These are right at the top of the list for most popular mods:

- **Power - Intake System:** A well designed intake can improve power and MPG by 5-10%.

- **Suspension - Lowering Springs:** These improve the appearance of your car and also make it handle better.

- **Styling - Wheels:** These immediately set your car apart from one just driven off the lot.

2. Does this void my warranty?

This is a loaded question. Generally, a dealership has to prove the mod caused the warranty problem. The Magnuson–Moss Warranty Act speaks to this fact. That doesn't mean that some dealerships won't try! It's recommended that you buy your Mazda from a mod friendly dealer or at least take your Mazda to one for service.

Even if one part of your warranty was nullified, that doesn't mean the entire warranty will be. Read the fine print, call around and talk to the dealerships in your area. You will find some more receptive than others and can determine what boundary's you feel most comfortable setting to maintain your factory warranty.

Ask! A good aftermarket parts company will give you an intelligent answer to this question and help you understand your rights should this issue ever come up. I have included a warranty icon on each of the mod cards in chapter 3 that will give you a general sense of how likely installing a particular part might affect your warranty.

Generally avoid 'extended warranties'. They often have clauses like "any non-stock' part voids it.

3. Can I install these parts myself?

You can save some serious cash by installing parts yourself. Many reputable vendors will have installation instructions posted online. Review them before you buy. Talk to the vendors support staff. If the staff gives you confidence, then you can do it!

Install instructions should give you a rating of how difficult the install is and let you know what tools you'd need to do it. Don't forget to get help from friends!

Some people prefer the peace of mind of a dealership or shop install. Ask the sales person before you buy the part how long it should take to install so an unscrupulous shop does not take advantage of your ignorance.

4. Do I need parts with CARB Approval?

California Air Resources Board (CARB) is a government agency in California responsible for improving air quality. Other states must either adopt the federal standards or CARB standards but cannot have their own. Over a dozen states have adopted the more aggressive CARB standard. Check with your local authorities (environmental protection agency) to see if the CARB rules apply to you. Enforcement will vary substantially by state.

Generally, CARB certification indicates that the modification will not negatively affect vehicle emissions according to CARB assessment and an executive order (EO) will be issued for the part approving it for use as legal in any CARB emissions standard state.

Example CARB EO#

Basically, any part that can affect engine performance is probably subject to CARB approval. A company that has received CARB approval can provide an EO number. The EO number should be visible on the part.

Unless you are in a state that is pro-active in enforcing a CARB policy, most people don't have to worry too much about enforcement. There are two exceptions.

- *First:* if your car gets a visual inspection during emissions testing. Many aftermarket parts do not make emissions worse and providing the part does not trigger a check engine light (CEL), it's possible to pass the sniff test, but a visual inspection may signal a red flag for the emissions testers.

- *Second:* if you break motor vehicle laws and get pulled over, the police can and often will look for violations during the traffic stop. Drive right and be respectful to police officers and you'll not likely have a problem.

5. Is Nitrous Safe?

With the popularity of movies like The Fast and the Furious, many people think they want to add a Nitrous bottle to their car. It can be an inexpensive way to provide a big boost in power. However, it can be very dangerous and cause serious damage to your vehicle if not installed or used properly. In over twenty years of modifying cars, it's not something I have ever opted to use in a build. It's far too easy to blow up an engine which can cost $5,000 or more to replace. Another downside is that you lose the extra power when your bottle runs out. I'd rather do mods that are permanent, require little maintenance, and are safe. Nitrous has its place for racers pushing the limit but not for your average street car.

6. Can I Get Sponsored?

If you are serious about getting parts for free or even at a discount you need to offer something in exchange. When companies choose to sponsor customers they are choosing brand ambassadors. How will you generate profitable customers or drive brand awareness? You need to have a solid answer to this question and probably a portfolio to submit to the company's marketing staff. It's not enough to offer to put some vinyl decals on your car and tell your friends how awesome your sponsor is. Tell the company exactly how you can help them. Be as specific as possible.

7. Do I need a tune? When is it helpful?

Typically, you can have about three power mods on a turbo car before a tune is recommended. With a naturally aspirated Mazda you can do a few more mods before moving to a custom tune. The factory fuel systems have a lot of margin for error and can handle modifications that increase power by 10% or more. A tune can help get more power but is often expensive and requires an expert with a dynamometer to do it right. Power programmers can help but are often expensive as well.

8. What's a BOV and a wastegate?

First, let's explain briefly how a turbocharger works. The exhaust gases flow past a turbine (wheel) with a compressor (wheel) on the other end of a connecting shaft. The energy of the exhaust gas spins the turbine and thus compresses the air in the intake side of the engine. The engine can generate more power with more (dense) air.

A blow-off valve (BOV) or bypass valve (BPV) allows the pressure in the intake side of the engine to dissipate when you let off the throttle so it does not damage the turbocharger. This extends the life of your engine and turbocharger. A BOV vents the pressure outside the intake tract (engine bay) while a BPV vents the pressure into the intake tract after the engines air sensor.

A wastegate is used to control the maximum pressure generated by a turbocharger which keeps the turbocharger operating in a safe range. The wastegate opens and closes, often many times a second, to keep boost pressure on target. Without a properly functioning wastegate, your turbocharger and engine life would be reduced.

9. How Can I learn What an Aftermarket Exhaust Sounds Like?

The best method, aside from finding a car with the exhaust already on it, is to look for video's on YouTube. You'll want speakers with a sub-woofer so that low notes will be reproduced otherwise all exhausts will sound just like that Honda with the $40 eBay muffler. All Mazda engines sound great with a properly tuned exhaust system.

10. Why is shipping so much?

You can sometimes find 'free' shipping on smaller parts. It's often difficult to do so for larger parts because the shipping cost is expensive relative to the cost of the part. For companies that offer 'free' shipping, many have built the shipping cost into the cost of their part. Exhaust systems, body parts, skid trays, and many other parts can be very large. All shipping companies use what's called dimensional weight to calculate shipping cost (https://wwwapps.ups.com/ctc). Below is a rough range of ground shipping costs by product type within the continental US. Express and international shipping will be 50-100% or more expensive on average.

PART	SHIPPING PRICE
Intake System	\$10 - \$20
Exhaust System	\$40 - \$80
Front Bumper	\$125 - \$250
Intercooler/Radiator	\$25 - \$75

Bonus: When should I upgrade the fuel pump in Mazdaspeed3/6 & Cx7?

While there is some variation from pump to pump, normally up to three power mods are safe on the factory fuel map from my experience. Once you upgrade the fuel pump or add a tune, you'll likely need both. The combination of these two parts can be over \$1,000. You can be confident your car is safe with mods by purchasing a Bluetooth OBDII connector (~\$20) and using free OBDII software on your smartphone (see mod card in chapter 3 for more details) to monitor fuel pressure.

US Market Versus The Rest Of The Planet

In this day and age people can buy parts from all over the world. Mazda models have several similarities across the different countries but also some key differences.

- **Bumpers.** Crash testing standards are different in various countries which means the bumpers are subject to different regulations. Some bumpers are shorter than the North American market models. This can affect body part fitment and the location and or fitment of the exhaust. If it looks different make sure you ask before you buy. Returning parts internationally is very expensive.

- **Power.** As long as the engine size and model year is the same then engine parts will usually work across the various markets. The US market cars often have larger engines than most of the rest of the world, so double check the engine size and not just the model.

- **Suspension.** Pay close attention to vehicle trim levels. Different size wheels and engines are good indicators that the suspension is upgraded. One problem I see over and over is with brakes. Be extremely careful buying brake rotors and pads across markets as these parts are often very different.

- **Styling.** This is another area where vehicle sub-models can be important. Even if the cars look the same, the parts may not fit if the trim has been upgraded

If you are going to purchase parts for your Mazda from another market be sure to consider at least these three things. First, make sure your sales person thinks it will work. Second, keep in mind there are often tariffs and brokerage fees that will be applied by the government and collected by the shipping company when bringing in parts from overseas that may not show up in the initial shipping quote, so don't be surprised when that gets tacked on by your local government authorities post purchase. Third, remember if you have to return parts because they don't fit, it could be very expensive to ship them back.

Lowering Your Car

There are two great reasons to lower your car; it looks great and it makes the car handle better. The downside is that your Mazdas underbody can scrape in steep driveways and your front bumper and doors are more likely to hit high curbs. This problem only gets worse the lower you go.

Go in/out of driveways at the sharpest angle possible. And if you are really a car lover, consider choosing your next home based upon how steep the driveway is!

Removing Catalytic Convertors

Catalytic convertors (cats) clean your exhaust to reduce impact on the environment. Removing them is bad for the environment and is illegal. It's almost never allowed on a street car. Cats do wear out, so when it's time to replace one you can look to the aftermarket for a higher flowing version. Some Mazda's can gain 10% or more horsepower by installing a high flow cat and even more if your original cat is failing.

Lighting Upgrades

These are very popular, especially with the advent of LEDs. However, your lights are an important safety feature of your car. The Department of Transportation (DOT) has strict regulations applying to certain vehicle lighting like headlights and taillights. Many overseas manufactured lights may not have DOT approval, but often Mazda Genuine Parts from European and Asian Mazdas will have different lights that are safe to purchase, though they can be expensive.

There are a lot of poor quality lighting upgrades available on the market. They leak, the plastic discolors, and the light is not very bright. I would recommend buying lighting products from reputable companies that honor warranties and allow returns for defective parts.

Careless Product Engineering

Any Joe Blow can make a part for a car. This doesn't mean they have the resources to do it properly. Watch for these signs of careless product engineering:

- **Bigger is better.** You left your BIB on your highchair when you were two years old. There is nearly always a point where bigger stops giving meaningful results and starts causing real problems.

- **Lack of testing.** Has the part been tested enough to ensure it will last as long as your Mazda? Don't let a lack of testing leave you stranded waiting for the tow truck.

- **Bolt-On.** This type of BO doesn't smell. Does the part install without permanent modifications to the car? You will probably want to return your Mazda to stock before you sell it so be sure to ask if the part requires permanent modifications to install and make sure to save any of the original equipment manufacturer (OEM) parts you remove in case you want to return the car to stock.

- **More parts required.** Make sure to ask if more parts are required. You don't want to be in the middle of a midnight install only to find that the part you purchased only works with other parts that company offers. Reputable companies build parts that work with stock parts and competitors parts whenever possible.

Value

Let's face it, there are lots of parts out there that do almost nothing for your average performance enthusiast. The parts in this book are 'bang for the buck' mods. Many of the parts not listed are either very difficult to install, require many other parts in order to get a real benefit from, or offer little in the way of improvement and so have very poor value for the beginner.

Stupid Upgrades Ruin Engines

If you decide to go crazy with your mods make sure you have a trusted expert to guide you. Once you commit to going for power mods outside the scope of this book the chances of blowing up your engine skyrocket. Not to scare you, because you can safely double the power output of your Mazda with proper guidance and probably another $5,000-$10,000. As a rule of thumb, remember the 10% rule. Seek expert guidance from reputable companies if you want to go more than 10-20% over OEM power.

2.6 Balancing Your Mods

Have you ever seen that guy at the gym with giant arms and then legs that resemble a chicken? If you spend all your time working on your upper body but leave out what actually holds your body up you will end up looking like Popeye. The same thing goes for cars. If you spend all your effort making your car go fast, but neglect handling or you significantly improve handling but forget about braking you will end up with an out of balance build. All the power in the world has nowhere to go without the right suspension and enough tire (aka traction) to get the power to the ground.

When you begin modifying your car, you will likely want to focus on power, handling or styling initially, but each has an impact on the other so it is best to take a balanced approach and incorporate some of each of these components into your build as you go along.

Power

All the suspension in the world won't give you the acceleration needed to get in front of that guy going 45 mph when you try to merge onto the freeway, you need power and quick acceleration for that.

If you overdo your styling components without considering what's under the hood, you may find yourself all show and no go. You can look good but if there is nothing impressive under the hood, people will quickly see that you have nothing to back up those looks.

Some of the best handling mods are wheels and brakes. Limiting weight that has to move up and down will increase grip and road feel. This is called unsprung weight. Imagine holding your arm out and hanging onto a weight. Now move it up and down. Not the easiest task. Now drop the weight and try this again with nothing in your hand. A whole lot easier. The same idea of weight transfers to how your car handles. The wheel and brakes are the weight.

Suspension

Everyone wants to go fast but what if you don't have the stopping power to compensate for your speed? You could find yourself in a compromising position. The right pads and rotors can make going fast safer and more fun. There is nothing quite like going 60 mph down a mountain and realizing the brakes have overheated and no longer work. You can't be the fastest if you can't make the next corner.

Suspension plays a powerful role in getting power to the ground. 500 horsepower will not translate to a fast car if you are just spinning your wheels all day long. Spring rates and dampeners play a huge role in getting power to the ground where you want it. Too much weight transfer under acceleration can unload the tires on a front wheel driver car resulting in excess wheel spin. Excess body roll during cornering can result in the inside tires slipping or spinning from decreased traction.

Styling

Handling and styling can go hand in hand. Downforce is a powerful tool to keep the car on the road or track, at high speeds. Useful aerodynamics can improve your handling and make your car look sleek, sexy and fast even when it's at a full stop.

Some styling mods can actually make your car perform worse. 24" wheels will slow your roll.

2.7 Terminology:

As with any hobby there are some specific acronyms and slang used within the community. In this book so far I have introduced terms like MPG (miles per gallon), and SRI (Short Ram Intake) which you may or may not already be familiar with. When looking into modifications there are a wide range of terms which can be used to describe a part in a car. Take the computer which runs the engine. It is commonly referred to as the ECU, CPU, computer, processor, chip, or brain box, just to name a few. I break down some of the most common acronyms and terminology in the glossary.

The glossary is by no means a complete list of terms, but it is does provide a good foundation for easier communication with other likeminded Mazda enthusiasts and will help you along as you dive deeper into discussing the various types of modifications you should consider for your Mazda.

CHAPTER THREE

HOW TO MODIFY YOUR MAZDA

This chapter starts with three key categories of parts to consider when modifying your Mazda: Power, handling and styling. I'm going to talk about what kinds of mods are available on the market for each of these categories, how to upgrade in stages to achieve the best possible build, and how well these parts play with each other. I will be introducing you to "Mod Cards" and will show you how to best utilize them. These will be a useful tool in helping you successfully modify your Mazda.

In the Power section I'm going to cover in detail the most common power mods on the market starting with your intake system all the way through to your exhaust. I'll touch on many things in between such as transmission mods, motor mount inserts, tunes, and oil catch cans.

After the Power section it's onto the Handling section. I'll talk about the benefits of lowering springs as well as coilovers, strut tower bars and also sway bars. Another popular modification I put in this section is in the braking department.

Next I'll cover the most visible modifications to your Mazda, styling. There are more options for styling than I could possibly list here so I'll stick with some of the most common. This is the section that will cover some of the ways to make your car look a lot better which includes, wheels and tires. Wheels and tires have tremendous benefits from a handling perspective, but most purchase upgraded wheels and tires for the style it adds to their Mazda, so it seemed most suited in this section of the mod cards.

Finally, I will provide you with staged mod cards that will outline how best to package your power and handling builds. This is a recommendation and by no means should be considered the only way to build out your Mazda. From my experience, the parts I list on each of the stages will play best with one another and provide the most benefits when installed simultaneously. I would highly recommend starting with the stage I power parts and then consider adding the stage I handling parts before moving onto stage II power parts (remember to balance your mods).

3.1 The Key To Understanding the Mod Cards

The mod card section is your quick reference guide for each of the various modifications I am recommending for you. This is an overview of the products with information about what they do, what they replace and what to look for when you are considering purchasing one. I have included icons to the right of all of the mod cards. These icons are meant to help everyone understand what the part does and how it will help you. About half of the icons are rated from one to three and are color coded. The best way to understand the rating is to understand what your ability is. Everyone can learn to do any of the modifications but starting with the hardest install might detour even the best of us. Basically, green is for the person that has never installed anything before. Yellow is for the person that installed a few things but might not be comfortable doing red activities. Red activities are the difficult or lengthy installs that will require more time, patience and probably a decent set of hand tools. All of the rest of the icons are explained in the guide on the next pages.

MOD CARD ICON GUIDE	
Time Estimate to Complete *This shows the relative time that it will take to complete the install. (1. Short, 2. Moderate, 3. Long)*	
Wrench Rating *This icon rates the skill level needed of the person installing the parts.(1. Little to no experience, 2. Has moderate experience with installation of parts, 3. Experienced)*	
Cost of the Part *This icon shows you how much you will be spending on this part.(1. Inexpensive <$200, 2. Mid-priced $200 - $500 , 3. More Expensive. >$500)*	
Warranty Risk *This icon rates the risk that you will void your warranty. (1. Little to no risk, 2. Has moderate risk, 3. Will almost certainly void your warranty.)*	
Power Gains *This icon rates the amount of expected horsepower gains. (1. Little gains <10hp, 2. Has power increase 10-30hp, 3. High power gains >30hp)*	
Handling Modification *This is a part that will help your Mazda turn better, stop better or react faster*	

MOD CARD ICON GUIDE	
Extends the life of the car *This type of part will reduce emissions or generally help to make your Mazda last longer.*	
Turbo Only *This part is for turbo only vehicles. This will not work on a non-turbocharged Mazda.*	
Upgrade Icon *This icon is for parts that are upgradable or changeable. Such things as multiple settings or ways you can configure the part.*	
Environmentally friendly *This icon is for parts that will decrease your carbon foot print or help make your engine cleaner.*	
Compliment *This icon is for parts that compliment other parts. You should look out for their mates or purchase them at the same time.*	
Fake Parts *This icon is for parts were there are lots of fake parts out there and you should watch out for poor quality or poor re-sults, and copy cats.*	

POWER

POWER

3.2.1 Intake

Upgrading your factory intake is the first and best power mod you can make to your Mazda. It's got great bang-for-the-buck. Most kits can be installed by a novice with basic tools if it comes with quality installation instructions. There are three primary types of intakes to consider.

By far the cheapest and easiest is a drop-in performance filter. I can't think of any reason not to at least upgrade your filter if you do absolutely nothing else to your Mazda. It takes five minutes to install and your original filter will inevitably have to be replaced. Since you will be replacing it anyway, you may as well get a high performance filter. If you don't know where your filter is located on your Mazda, most owners' manuals will show you how to replace it or check out Appendix A.

Removing the filter and holding it up to a light will show you how dirty it is. If it's really caked with dirt, it obviously needs to be replaced. Trying to shake or blow the dirt out is a waste of time because too much of it will be embedded in the filter fibers.

A better option to upgrading your filter is a short-ram intake (SRI). This type of intake offers the benefit of a shorter, smoother path for air to travel to enter the engine (think less resistance). These intakes have a high-flow conical filter that sits in the engine bay. Their high placement in the engine bay is optimal for keeping your filter dry. The downside with this filter position is that you can temporarily loose some power if the vehicle is not moving, because heat builds up under the hood. This results in the engine drawing in less dense air and will make your car feel a little sluggish when you start it to drive away after a stop. However, after a short drive, the cooler air from outside will cool the engine bay down. As an optional addition, many people chose to add a Cold Air Box. This acts as barrier between the filter and the hot air in the engine bay and will help prevent some of the temporary loss of power I just referred to.

A Cold Air Intake (CAI) offers similar power gains to a short-ram intake with the added benefit of constantly drawing cooler air. Because the air has a longer path, the power gains are comparable to a SRI. The primary difference between a CAI and SRI is that the CAI does not suffer as badly from heat soak.

Myth Busted

Running through a deep puddle will not 'hydrolock' an engine equipped with a CAI. A little water can actually be beneficial to your engine as it helps clean internals. However, if you drive your car into a lake, you could have a problem. If you are expecting a flood or hurricane to pass through leave the CAI equipped Mazda at home and you'll be just fine.

TIP

Additional horsepower can be gained by replacing the turbo inlet pipe. Look for intake options that include a turbo inlet pipe for maximum gains.

SHORT-RAM INTAKE (SRI)

What it Replaces

The factory air box, filter, and intake tubing

Why You Should Want It

- *It sounds awesome*
- *Increases horsepower and torque*
- *Increases gas mileage*
- *Shortest air path*
- *Great bang for your buck mod*

What You Should Look For

- *Oil-less air filter that can be cleaned and reused*
- *Multi-layer reinforced silicone connectors*
- *Precision machined MAF housing to avoid CEL*
- *Some cars require an air straightener*
- *CARB certification (if you need it)*

COLD AIR INTAKE (CAI)

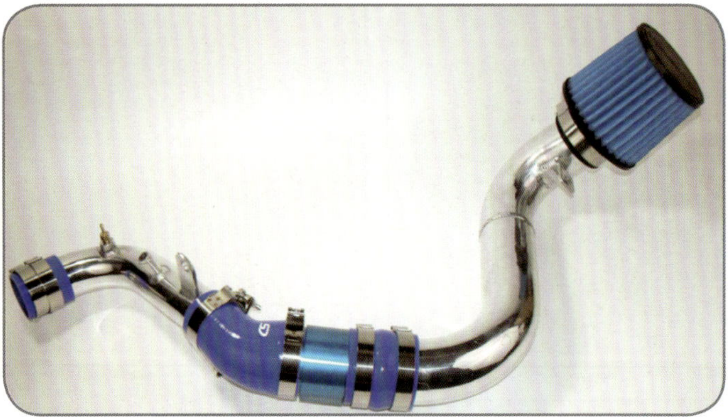

What it Replaces

The factory air box, filter, and intake tubing

Why You Should Want It

- It sounds awesome
- Increases horsepower and torque
- Increases gas mileage
- Lots of fresh cool air for your engine
- Great bang for your buck mod

What You Should Look For

- Oil-less air filter that can be cleaned and reused
- Multi-layer reinforced silicone connectors
- Precision machined MAF housing to avoid CEL
- Some cars require an air straightener
- CARB certification, only if you need it

COLD AIR BOX

What it Replaces

Adds a box around your short ram intake to shield it from the hot underhood air of your engine bay

Why You Should Want It

- *Easy to install*
- *Will help to get the most HP and torque from your SRI*
- *Reduced lag off of a stop from heat soak*
- *Keeps your filter cleaner longer*

What You Should Look For

- *Temperature testing*
- *Durable, lightweight construction*
- *Must be designed specifically for your brand SRI*

DROP IN FILTER

What it Replaces

The factory air box filter

Why You Should Want It

- *Easy to install*
- *Increases horsepower and torque*
- *Quiet*
- *You will have to replace your air filter any-way*

What You Should Look For

- *Oil-less filter*
- *Reusable*
- *Water resistant*

3.2.2 Exhaust

Upgrading your exhaust can result in modest power increases on naturally aspirated (NA) cars and big power increases on turbocharged cars. As a result, I wouldn't call an exhaust upgrade a bang-for-the-buck mod for most NA cars but I would for turbo cars. That said, I'd upgrade the exhaust on any car I own! A properly tuned exhaust system will sound great and free up power lost to the factory mufflers and pinch-bent piping.

ENGINE TYPE	FRONT	MIDDLE	BACK
Naturally Aspirated	Header	Main Catalyst (2nd Cat)	Cat Back Exhaust
Turbocharged	Downpipe	Main Catalyst	Cat Back Exhaust

Exhaust systems can be broken into three components: 1. Downpipe, 2. Main Catalyst, 3. Cat Back Exhaust

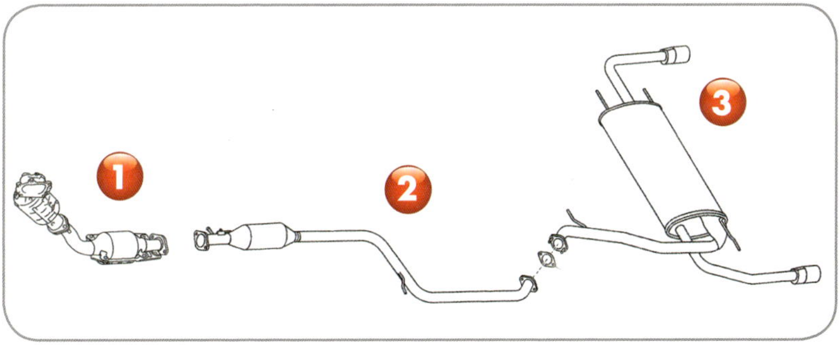

Up front, a header can give 0-10 HP but a quality header can be expensive so I don't recommend it as an entry level bang-for-the-buck mod. However, a downpipe can free up all kinds of power in a turbocharged car.

In the middle, you'll find the primary catalytic convertor. Big gains can be had by upgrading the main catalyst to a high flow cat, but since this 'filter' protects our environment and is very heavily regulated by our government, I recommend people don't replace it on street cars until the factory catalytic convertor fails or gets damaged. Once that happens, look for a high flow option from the aftermarket. Some people like to run a pipe to delete the catalytic on the race track. These are called racepipes or test pipes. These should be installed at the track and removed after racing before driving on public roadways.

The cat-back exhaust includes a section that runs from near the mid-point of the car to the rear axle and a section that runs from the rear axle to the rear bumper. Cat-back exhaust will yield 5-10% horsepower on most late model Mazda's.

CAT-BACK EXHAUST

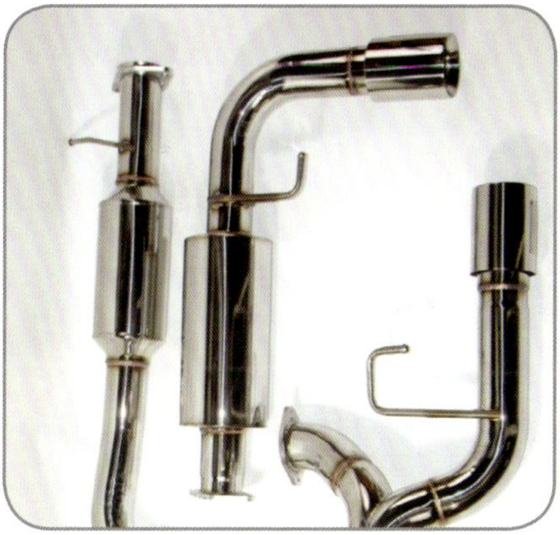

What it Replaces

The entire exhaust system after the 2nd cat., includes piping as well as a resonator and muffler(s)

Why You Should Want It

- It sounds awesome
- Increases horsepower and torque
- Should not affect warranty
- Looks great (bigger exhaust tip)

What You Should Look For

- Stainless steel piping, flanges, and muffler packing
- OEM style flange to catalytic convertor
- Mandrel bent piping with TIG welds
- All necessary hardware and gaskets

DOWNPIPE

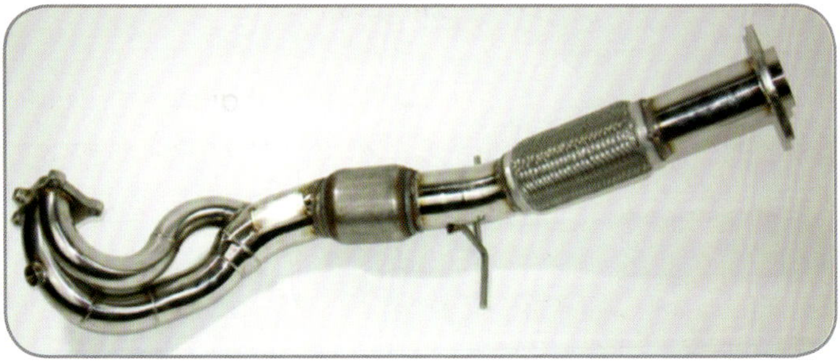

What it Replaces

The factory exhaust on a turbo car from the turbo manifold to after the catalytic convertor

Why You Should Want It

- Increases horsepower and torque
- Frees up exhaust flow
- Good replacement part if your factory cat is beginning to fail

What You Should Look For

- High flow catalytic convertor
- Stainless steel piping, flex pipe and flanges
- OEM style flange to the rest of the exhaust
- Mandrel bent piping with TIG welds
- All necessary hardware and gaskets
- Good instructions (this is a tricky part to install)

3.2.3 Transmission and Drivetrain

There are many ways you can modify your transmission and drivetrain but I'm only going to talk about the entry level mods that are most common, reasonably priced, and will add some enjoyment to your driving.

The first Transmission mods I'd like to talk about are Short Shifters. Short Shifters attach to either the actual shifter inside the passenger compartment of the car or to the transmission itself and can be anything from a simple bracket that changes the geometry of the shift linkages or can be a full replacement shifter assembly. By reducing the distance the actual shifter has to travel from one gear to the next you reduce the amount of time taken to make each shift. Less time spent shifting means more time spent putting power to the ground.

The downside to this is that adding a short shifter can make it harder to shift. Most people would agree that the benefits far outweigh the negatives with a short shifter and you will enjoy more spirited driving once you have added one.

Another great "transmission" upgrade is engine mounts. These are often overlooked but on many front wheel drive vehicles, the factory mounts can cause what is called wheel hop (a bouncing feeling when a front wheel drive car loses traction) and torque steer (the feeling of the car pulling you left and right as you accelerate). You have a couple options for upgrading your engine mounts. You can add inserts which will fill in the gaps in the factory mount without adding as much of the NVH (noise, vibration, and harshness) commonly associated with aftermarket motor mount replacements.

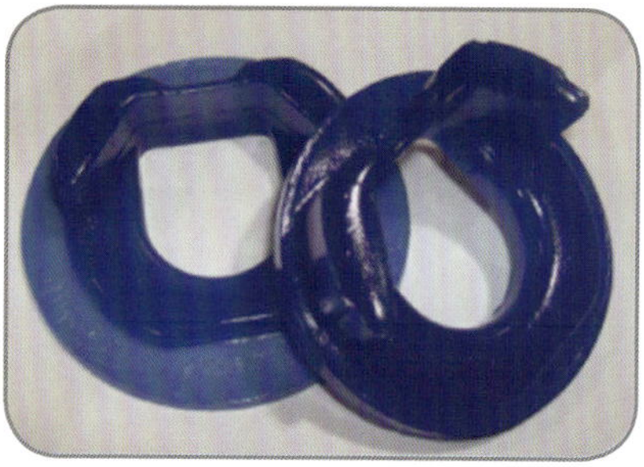

The other option is a full motor mount replacement. From the factory, motor mounts are designed to support the engine with rubber bushings. The natural flexibility in rubber allows the mounts to absorb vibration, but also allows the engine to move, a lot. By replacing them with a stiffer material like urethane you will no doubt feel the improved handling, but the trade-off is you will also feel more vibration from the engine and the sound will transfer more directly into your cabin.

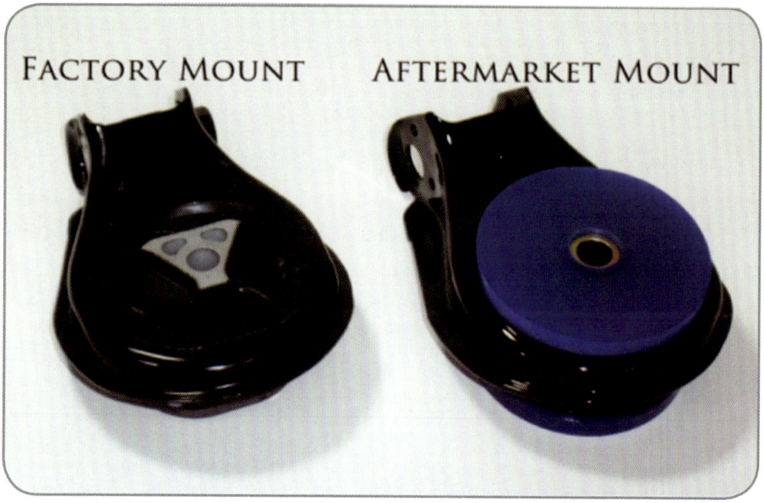

SHORT SHIFTER

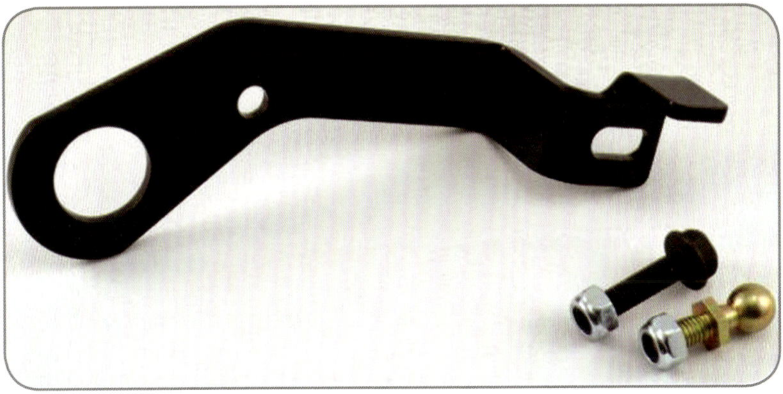

What it Does:

Reduces the travel of the shifter forward and backward, this allows for quicker shifting

Why You Should Want It

- *Reduces wasted time in between shifts*
- *Increased feel and action*

What You Should Look For

- *Proven product*
- *20 - 30% reduction in shift throw*
- *Easy installation*

Considerations

- *Adding a heavier shift knob will make shifting easier*

ENGINE MOUNT

What it Replaces:

The factory soft rubber engine mounts that allow a lot of engine movement

Why You Should Want It

- *Quicker throttle response*
- *Will last much longer than stock*
- *Improved handling*
- *To replace a worn mount*

What You Should Look For

- *Polyurethane*
- *Light weight*
- *Quality construction*
- *Reasonable noise vibration and harshness (NVH)*

Other Considerations

There are various levels of stiffness of polyurethane available on the market. As a rule of thumb 70 durometer is for a daily driver and 80 or more is good for the track.

3.2.4 Other Popular Power Items

Now that we covered the basic power and drivetrain mods, I will touch on some other under the hood modifications that are very popular but aren't necessarily what people may think of when they think engine.

Oil Catch Can (OCC)

This is literally a can that catches oil, water vapor and other emissions in your crank and cam case. As the engine runs there is a small amount of compression (gases and fuel) from the cylinders that leak into the crank and cam case. This vapor mixes with oil and contaminants in the engine end up building up in your intake. The can catches them before they re-enter your engine leaving the intake manifold clean and reducing buildup on the valves

Once you have installed a catch can, be sure to monitor the level of fluid it is catching and empty it regularly, especially at first. If you allow it to become too full, it will have a negative effect by allowing the engine to suck the contaminates contained in the can back in.

Tuning (Reflash)

Your ECU (Engine Control Unit) has mapping that allows fuel and spark to be delivered at a certain period of time. A reflash can change fueling, ignition timing and even turbo boost pressure. There are several out-of-the-box reflashing units on the market.

Blue Tooth Logging

There are many options now for "logging your car". You can get inexpensive blue tooth OBDII adapters on Ebay that will allow you to log your engine parameters, clear check engine lights, and can help you maintain your Mazda right from your smart phone. I recommend the Torque Ap.

Fuel Pump Internals

Once you have added three or more engine modifications to your direct injected Mazda, you might notice your fuel pressure beginning to drop. The best way to solve this is by upgrading to larger internals in the pump itself. This allows pressure to stay where it needs to be for safe engine operation. If you monitor fuel pressure with blue tooth logging, you can decide when your car is ready for this upgrade rather than letting someone tell you that you need it, when it is possible you don't.

OIL CATCH CAN

What it Replaces:

Nothing, it is an additional filter

Why You Should Want It

- *Keeps oil out of your intake system*
- *Maintains cleaner valves for longer engine life*
- *Minimizes burning PCV system oil and vapors.*
- *Better emissions*

What You Should Look For

- *Easy to drain*
- *Easy to view contents*
- *Solid test results*
- *Complete kits for easy installation*

TUNING (REFLASH)

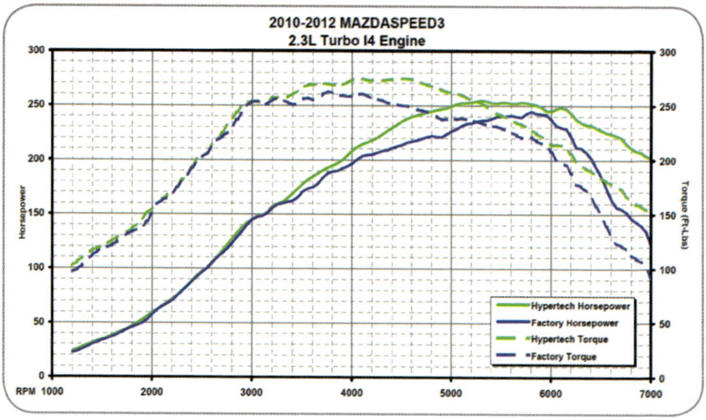

What it Replaces:

The existing ECU mapping with a performance map for more power

Why You Should Want It

- *Increased power*
- *Better MPG*
- *Can optimize for your climate*

What You Should Look For

- *Plug and play with your car*
- *Reputable brands*
- *Pre-existing maps that are coded for your mods*
- *Lots of support for that tuning device from custom tuners*

BLUETOOTH LOGGING

What it Replaces:

Nothing, adds ability to monitor vehicle systems

Why You Should Want It

- *Logging*
- *Clearing check engine lights*
- *Troubleshooting issues with your car*
- *Get real time information about your car*
- *Test results and graphing*
- *Monitor fuel pressure*

What You Should Look For

- *Bluetooth OBDII adapter*
- *Free or cheap software*
- *Faster logging speeds*
- *Plug and play*

FUEL PUMP INTERNALS

What it Replaces:

The factory high pressure fuel pump internals with larger internals

Why You Should Want It

- Keeps fuel pressure from dropping
- More head room for future mods

What You Should Look For

- Good installation instructions
- Solid testing results from well established company
- Replacement spring
- Complete kit

3.2.5 Turbo System Parts

All modern Mazda turbo vehicles have components that are easy to change and will make a dramatic increase in performance. The two most common and easy to swap out include your blow off valve and your top mount intercooler. Though there are other options on the market but they are more advanced and expensive, so I will cover them later in "Taking your mods to the next level" and just stick with what I believe to be the most beneficial, easiest to install items for now.

Blow off Valves

The most common question I hear is, "what does it sound like?" since this is the component that creates the "wooshing" sound you hear when you shift between gears. Imagine your turbo is a giant hairdryer pushing out hot air. In between shifts the throttle plate shuts but the turbo is still spinning and compressing air. The air has no place to go when the throttle plate shuts causing the hot air to travel backwards towards the turbo. This can create turbulence and can stall the compressor wheel causing seal and bearing failures. The blow off valve or bypass valve releases that hot air pressure.

There are two primary kinds of BOV/BPV

- **Recirculating** – Most cars have a valve that returns the compressed air from the turbo system back into the intake when you shift. This is known as a recirculating valve. By recirculating the air, the mass air flow sensor is able to correctly keep track of engine air.

- **Vent to Atmosphere** – This valve actually vents to the engine bay. This can cause temporary incorrect readings of the mass air flow sensor due to the "escaped" air.

Top Mount Intercooler

An aftermarket Top Mount Intercooler (TMIC) is a direct replacement for the Mazda Stock Top Mount Intercooler. It is designed to provide improvements over the stock intercooler without breaking the bank. It's an easy to install item that will provide an excellent boost in power.

When looking for a quality aftermarket top mount intercooler, you should always look for one that has been flow tested and shows significant improvements in flow rates. Additional airflow means your Mazda will be able to handle more horsepower. You can expect a good Top Mount Intercooler to handle up to 400 hp without losing efficiency. You should also look for one that uses cast end tanks. Some intercooler designs lack flow to the extremities of the core causing the center of the core to be "overworked" and the outside edges to be "underworked", by using cast end tanks, an optimal flow is created across the entire width of the intercooler. This means a faster response time and an improvement in power.

BLOW OFF VALVES

What it Replaces:

The factory blow off valve or bypass valve

Why You Should Want It

- *Extend turbo life*
- *Improves the sound of your car between shifts*
- *Often needed to run higher boost pressure*

What You Should Look For

- *Quality manufacturer*
- *Sound clips*
- *High quality construction and material*
- *Fitment to OEM flange*
- *Watch out for size: clearance can be an issue*

TOP MOUNT INTERCOOLERS

What it Replaces:

The factory Intercooler with a more efficient, larger intercooler core

Why You Should Want It

- More power because of less airflow restrictions
- Less heat soak
- More consistent power
- Great bang for your buck mod
- Easy to install

What You Should Look For

- Proven efficiency
- Maintain factory shroud
- Bar and plate design
- Cast aluminum or other high quality construction material

HANDLING

3.3 Handling

3.3.1 Suspension

The suspension controls the contact your car has with the road. By upgrading key components, you will be able to improve handling and your overall driving experience. Struts, shocks, and springs are all essential components of your suspension system. I mention struts and shocks (dampers) here even though I am not including it as a bang-for-your-buck mod because they can vastly improve handling. Coil-overs have the dampers built in, so I feel they are a better value than upgrading springs and struts separately.

Have you ever dropped a spring and watched it bounce back up from the ground? Without the dampening effect of struts and shocks, the same thing would happen to your Mazda when you hit a bump. Lowering springs, especially really low ones can wear out your OEM struts and shocks faster.

For springs, there are a few options on the market that you can purchase to lower the center of gravity on your Mazda and improve the handling. I will focus on the two most popular.

- **Lowering Springs:** This popular modification is often times one of the first. That is probably because they are relatively inexpensive and make a world of difference. Adding lowering springs will not only reduce that ugly wheel gap between your tires and fender, but they will also provide great handling performance and a better road feel.

- **Coilovers:** This is a more expensive option because you are replacing your springs and your dampers in one package. Coilover kits match upgraded springs and dampers to provide ride height adjustment and dampening adjustment.

Coil-Over Compared to OEM Strut & Spring

TIP

As you lower your car your suspension geometry is changed. Get a 4 wheel alignment after installing lowering springs or coil-overs or you just might be buying new tires a lot sooner than you want to.

Be careful when choosing your suspension parts. Buying the wrong parts can create an uncomfortable ride, which will give you bone jarring hits over bumps in the road. A properly setup suspension will reward you with the precise handling in the corners and allow you to "dodge bugs".

LOWERING SPRINGS

What it Replaces:

Replaces the factory springs with ones that lower your Mazda

Why You Should Want It

- Just say no to the 4wd wheel gap!
- Lower the center of gravity for better handling
- More aggressive, sleeker look
- Less body roll and quicker turning response

What You Should Look For

- Superior construction
- Corrosion resistant coating
- 1"- 2" in drop best for daily drivers
- Firmer spring rate

COILOVERS

What it Replaces:

The springs and struts/shocks in your Mazda with a matched setup giving you the ability to fine-tune the suspension to your personal taste

Why You Should Want It

- *Adjustable ride height*
- *Lower center of gravity for better handling*
- *More aggressive, sleeker look*
- *Adjust ride quality to optimize the handling for the street or track*

What You Should Look For

- *Adjustable dampening without removal*
- *Corrosion resistant components*
- *Rebuildable dampers*
- *Adjustable camber plates inlcuded with set-up*
- *Helper springs for quiet operation*

3.3.2 Braces and Bars

A Mazda off the factory floor has been designed to provide a comfortable ride for the average Joe to absorb the roughness of the road so bumps and divots do not transfer from the frame into the steering wheel and seat. Mazda has designed the chassis to have some "give" or flex. The tradeoff is that when the chassis is pushed in hard cornering the car can be unstable and loose traction. Strategically placed reinforcements like frame braces and strut tower bars help maintain chassis and suspension geometry for optimal traction.

Strut Tower Bar

The tower bar sits between the two shock towers. By connecting these two points in the vehicle you will be able to reduce vehicle chassis deformation and improve cornering. Most cars have front and rear strut tower bars available.

2011 Mazda 2

Chassis Brace

On the unibody construction of most modern cars there are sections of stamped sheet metal which are welded and bonded together. Providing extra bracing to the stock unibody reduces the twisting of the chassis. The more common areas to brace are near the suspension mounting points or in large open spaces. A good example of the suspension mounting points is the front suspension control arms of the Mazda3/ Mazdaspeed 3.

The front mounting points for the arms on the vehicle are two pieces of the sub-frame which extend forward with no lateral support.

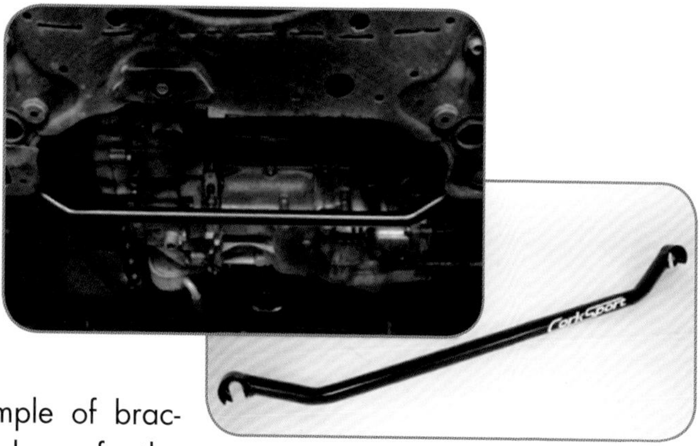

By adding a brace between the two points, the mounting points are strengthened to decrease the amount of flex in the body. An open space example of bracing is a rear hatch brace for the Mazdaspeed 3. If you have seen a Mazdaspeed 3 with the seats down there is a huge open space. This is great for hauling cargo but not the best thing for keeping the chassis from flexing under cornering. The solution is to brace the rear area from the left to right suspension points. Ideally you want a brace which can be removed easily, in case you need to help your hot neighbor move something large in a hurry.

Sway Bars

Your Mazda already has a sway bar (also known as an anti-roll bar or anti-sway bar) on it. It connects the opposite wheels together and helps to reduce the roll (weight transfer) during cornering. Aftermarket sway bars are usually larger than the factory bar which increases weight transfer from the inside to the outside for better traction. Often times you can purchase adjustable sway bars that will allow you fine tune your suspension even further and dial in understeer/oversteer to your driving preference.

STRUT TOWER BARS

What it Replaces:

Nothing, adds Bracing

Why You Should Want It

- Reduces chassis flex
- Increases cornering ability
- Great bang for your buck handling mod
- Easy to install

What You Should Look For

- Aluminum construction for minimum weight
- Watch out for size, clearance of engine and hood can be an issue

SWAY BARS

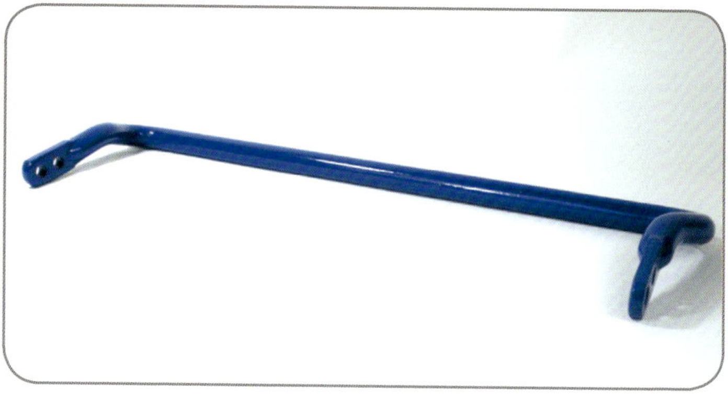

What it Replaces:

Replaces the factory anti-roll bar to provide more balance while cornering

Why You Should Want It

- *Increased car control during cornering without the decrease in ride quality typical of other mods*
- *Adjustable bars allow you to fine tune to under steer/over steer to your liking*

What You Should Look For

- *Corrosion resistant coating*
- *Includes upgraded greasable bushings*
- *Published specifications for increased stiffness*

3.3.3 Brake Parts

Brakes are often the last part upgraded on a car, but shouldn't be. Being able to stop isn't just important it is critical to keeping you safe while driving.

Pads

This is a part that will need to be replaced periodically over the life of your vehicle. Since you will be purchasing them anyway, it's a great opportunity to upgrade to performance brake pads. So which pad is best for you? First, you want to look at the material the pads are made from. There are ceramics, semi-metallic, organic and metallic formulations. For daily driving (street use), ceramic and organic compound brake pads work great because they do not require "warming up" like other compounds do. They also create less brake dust and noise so you won't dirty up your new wheels as fast. On the other hand, if you are looking for something that will perform best on the track, you may want to consider semi-metallic or metallic brake pads. These pads need to warm up to perform to their full potential, but once warm, offer superior braking power. A downside for metallic pads is that they create quite a bit of brake dust and noise.

Rotors

Brake rotors are on the receiving end of the brake pads. The brake pad composition as well as your driving style will determine how long your rotors last. The ultimate goal in braking is to stop your car. Components that eliminate the most amount of heat will help to extend the amount of time you can use your brakes before they overheat and stop working effectively. There are several different varieties of upgraded versions of rotors. Slotted/vented rotors allow for more efficient heat dissipation which reduces brake fade. Cross-drilled/vented rotors are similar to the slotted/vented but are not recommended for racing applications due to the possibility of the rotors cracking between drilled holes. Cryo-treated rotors increase the durability of the rotor further, so look for this option if you expect to abuse your brakes.

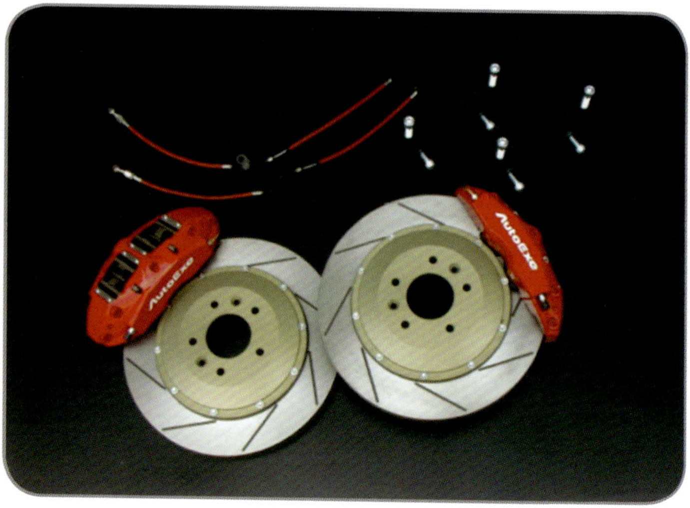

BRAKE PADS

What it Replaces:

Replaces the factory brake pads with compounds which can improve your braking distance

Why You Should Want It

- Shorter stopping distances
- Better resistance to fade under heavy use
- You have to replace them anyway, you may as well upgrade
- Less dust to keep your wheels clean

What You Should Look For

- Good description of recommended pad usage
- Non-toxic materials
- Good in all weather conditions
- Low noise levels

BRAKE ROTORS

What it Replaces:

The factory rotors with upgraded slotted/vented or cross-drill/vented rotors

Why You Should Want It

- *Greater heat dissipation*
- *Increased durability*
- *Less brake fade, better stopping*

What You Should Look For

- *Durable material*
- *Reputable manufacturer*
- *Testing results*
- *Meet or exceeds OE standards*

STYLING

STYLING

3.4 Styling

3.4.1 Tires

Content provided by Justin Metz – Goodyear Tire

As the only point of contact between your car and the road ahead, few upgrades can instantly improve vehicle performance like a new set of tires. However, it's important to note that all tires are not created equal. To get the most out of a tire upgrade, you'll need to choose a performance tire that matches your needs and driving style.

There are two main categories of high performance tires: summer and all-season. As the names suggest, these tires are defined by the type of weather they'll excel in. A high performance summer tire offers optimal traction in both dry and wet conditions, but is not intended to be driven on snow or ice. High performance all-season tires, on the other hand, may trade some dry traction to provide all-season versatility—even in winter weather.

If you live in an area where snow is not an issue, a high performance summer tire is the obvious choice for getting the best performance from your car. A wide variety of performance summer tires are available on the market, ranging in features and price. An example of a premium summer performance tire is the Goodyear Eagle F1 Asymmetric 2. This tire is an excellent choice for drivers looking to get the most out of their car in everyday driving. It features a tread compound with motorsport resins for superior dry handling and a lightweight, racing-derived construction for enhanced steering precision and handling.

Drivers looking for added performance on the track or autocross course may prefer something a little more extreme. Extreme performance summer tires, like Dunlop's new Direzza ZII, will trade some comfort, wet traction and tread life for even more dry grip—the perfect formula for competitive driving. The Direzza ZII is the successor to Dunlop's popular Direzza Sport Z1 Star Spec, and offers impressive traction for the serious driving enthusiast.

If you want the performance of a summer tire, but live in an area where you'll encounter snow and ice, you may want to consider purchasing a set of dedicated winter tires. With this option, you can have the best of both worlds by swapping between your summer and winter tires when the seasons change.

For those who want one tire to do it all, a high performance all-season tire may be the best option. These tires typically use all-season tread designs and rubber compounds to provide year-round traction, even in light snow. Despite their all-season capabilities, these tires are still more than capable of providing the handling and traction you'd expect from a high performance tire. If you're looking for a premium all-season high performance tire, consider the Goodyear Eagle F1 Asymmetric All-Season. This tire's specialized tread compound delivers traction in any season, and its asymmetric tread design features specific dry and all-season zones to combine race-inspired handling in warm weather with all-season traction in cold temperatures.

In conclusion, whether you're looking to build some confidence through the corners or shave seconds off your lap times, a tire upgrade is a great place to start.

SUMMER TIRES

What it Replaces:

Your vehicle's original equipment tires

Why You Should Want It

- *Enhanced grip and traction*
- *Optimal performance in dry and wet conditions*
- *Shorter stopping distance*
- *Faster lap times*
- *Lighter weight*

What You Should Look For

- *Quality construction*
- *Smaller side wall if you're plus sizing wheels*
- *Wet and dry traction*
- *Tread life*
- *Widest tire possible*

Other Considerations

Tires should always be mounted and balanced by a professional, consider upgrading your wheels at the same time

ALL-SEASON TIRES

What it Replaces:

Your vehicle's original equipment tires

Why You Should Want It

- *Made for driving in snow and ice*
- *Enhanced grip and traction*
- *Shorter stopping distance*
- *Faster lap times*

What You Should Look For

- *Quality Construction*
- *Smaller side wall if you're plus sizing wheels*
- *All season capabilities*
- *Wet and dry traction*
- *Tread life*

Other Considerations

Tires should always be mounted and balanced by a professional, consider upgrading your wheels at the same time

3.4.2 Wheels

Content Provided by Justin Banner – Mackin Industries

If there is only one upgrade you can do to your car that helps it stand out the most, it would be your car's wheels. Things to consider:

Diameter

The diameter of your wheels can actually change the acceleration and speed of your car. You can increase or decrease the un-sprung weight affecting the rotational inertia. Tire availability and sizing choices will differ depending on the diameter you are looking for. Keep in mind, the smaller the wheel, the smaller your brake package has to be in most cases.

When you go up sizes, it's known as "Plus Sizing." A +1. You can also "Minus Size," as a few autocross and road racers do to for added acceleration.

Width

Width will affect the weight and tire width you can use on your car. Like increasing the diameter, increasing the width will add more material which adds more weight and rotational mass. This may require more braking power unless the wheel is made of lighter materials or better construction. If you are "stretching" a tire you will go with a narrower tire with a wider wheel.

Offset

Offset is the measurement from the center of the wheel to the back of the wheel hub. A positive (+) offset moves the wheel hub closer to the wheel face, a zero offset is dead center, and a negative (-) offset is closer to the brake rotor.

Style

You can find everything from solid wheels, multiple spokes, wire wheels, and anything in-between. There can be some detriment in choosing certain styles of wheel faces in terms of mass and some have a lack of openings for brake heat to escape. Some wheel faces can also benefit your aerodynamics, brake heat extraction, and even wheel mass. Certain wheel faces can affect your brake package, so be sure to check that your caliper will clear the face of your wheel.

Wheel Mount Style

How the wheel mounts to the hub is critical as well. Many aftermarket wheels offer the ability to be mounted using a hub centric adapter. This will physically center the wheel by the center of the hub as you mount it. This will ensure you don't get wheel wobble as you drive down the road. There are wheels that are known as lug centric, which means the lug nuts center the wheel rather than the hub.

Wheel Construction

How the wheel is made is important in choosing the right set of wheels. Cast wheels offer a much lower price point since producing a cast wheel is far less involved than a forged wheel. A forged wheel will be stronger and lighter since it is made out of a single piece of metal.

Always do your research. When it comes to your wheels, tires, brakes, and suspension components, remember, these parts stop, turn, and suspend your car and if you have a failure with any of these at any speed, the results are never good.

FORGED WHEELS

What it Replaces:

Replace your OEM steel or aluminum wheels

Why You Should Want It

- *Lighter than most OEM and cast wheels*
- *Stronger than most OEM and cast wheels*
- *Higher reduction in unbalanced weight*
- *Better acceleration*
- *Will not affect your manufacturer's warranty*

What You Should Look For

- *JWL and/or TUV certification*
- *Manufacturer stamp or decal on the wheel*
- *Light weight*
- *Proper offset for fender and strut/shock clearance*

Other Considerations

Beware of cheaply made imitation wheels

CAST WHEELS

What it Replaces:

Replace your OEM Steel or aluminum wheels

Why You Should Want It

- Lighter than most OEM and cast wheels
- Stronger than most OEM and cast wheels
- Reduction in unbalanced weight
- Are available in the size and offset you want or need
- Will not affect your manufacturer's warranty

What You Should Look For

- JWL and/or TUV certification
- Manufacturer stamp or decal on the wheel
- Light weight

Other Considerations

Beware of cheaply made imitation wheels

3.4.3 Exterior Styling

Exterior styling is like the frosting on the cake for your Mazda. It is what will give your car character and will help other people see your swagger with a quick glance at your ride. Exterior modifications can range from the practical to the extreme and everything in-between.

Many people may remember the RX-7 from the Fast and Furious Tokyo Drift with the Veilside Fortune body kit that lives on as one of my all-time favorite movie cars. This is an example of taking exterior modifications to the extreme. To add a full body kit like that can easily cost you $20,000 plus dollars to purchase, install, and paint to match.

Really, the hope is that you bought the car because you already like the way it looks and you just want to add exterior modifications to enhance and personalize the vehicle to your taste. You can add simple modifications like LED's, tow-hook plates, antennas, lowering springs, window tint and spoilers to modestly affect the profile of your Mazda.

One of the more common modifications to newer Mazdas is a license plate relocation kit. There is nothing worse than having your friends check out your new car and seeing 2 big screw holes in your front bumper because the dealer installed the rather ugly mounting bracket. Mazda did not intend for it to be there when they did the styling of the car. The relocation kits set the plate off to the right or left of the opening and give the plate a more purposeful look.

One thing to remember when purchasing exterior parts, especially body parts, is to make sure they fit your Mazda before you install them. Most companies will not take something back if it is installed or modified from you trying to fit it.

LICENSE PLATE RELOCATION KIT

What it Replaces:

Moves your front license plate from the center of your bumper to the tow hook location

Why You Should Want It

- *Improves the look of your car from the front*
- *Allows more airflow under the hood*

What You Should Look For

- *Impact and temperature resistant backing plate*
- *Bumper plugs to fill holes*
- *Security fasteners to avoid plate theft*
- *Mounts to tow-hook screw*

Other Considerations

When you purchase your new car, it is a great idea to ask your dealer not to install the front plate. This way you will not have holes in your bumper.

MAZDA GENUINE ACCESSORIES

Alan Webb Mazda Vancouver, WA

There are plenty of options available at your local Mazda dealership. You can purchase a different style bumper or spoiler. They also have plenty of add-ons to choose from like roof racks, front masks, upgraded car doors, cargo boxes, rear bumper guards and oil caps. Mazda dealers have parts already matched to your paint color. This can save you a lot of money as custom paint is expensive.

MUD FLAPS

What it Replaces:

Adds mud flaps to the front and rear of your Mazda

Why You Should Want It

- *Protects your vehicle from road debris, rocks and the elements*
- *Helps to prevent paint chips*
- *Keeps the underside of your car cleaner*

What You Should Look For

- *High quality urethane that will not tear or fade*
- *Installation hardware*
- *Length suitable for your Mazdas ride height*

FOG LIGHTS

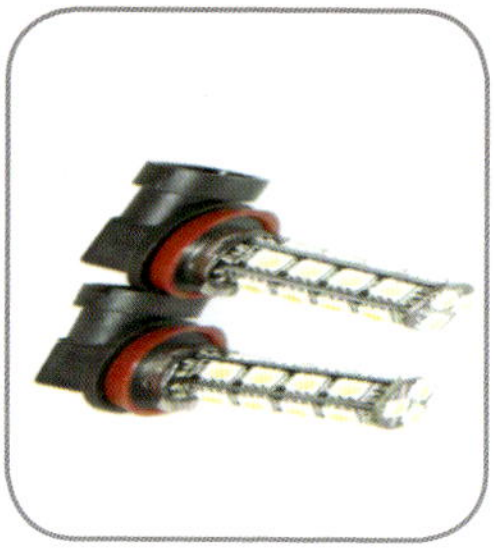

What it Replaces:

The incandescent fog lights with LED lights

Why You Should Want It

- *Improves the look of the exterior lighting*
- *Generates less heat*
- *Resists shock and vibration better*
- *Longer lifespan*

What You Should Look For

- *A kit that includes enough LED's to effectively work as a fog light*
- *Highly respected vendor to ensure quality*
- *Installation instructions*
- *No custom adapters or special wiring needed*
- *Color temperature/light color you like (6500 range is popular)*

Other Considerations

Should not be used for headlights due to the low power draw

3.4.4 Interior Styling

Now this is where form and function collide. Interior styling parts can boost the look of your Mazda and have practical performance advantages. Modifications for the interior of your vehicle include audio components and interior lighting among others. You may not consider switching out your stereo (head unit) right away but might want to consider adding amps, bass and speakers for a crisp clean sound. Once you have had your vehicle for a few years, there are bound to be some new and exciting technologies out on the market that might entice you to upgrade the stereo. Dealer navigation can be part of a $2,000 or more tech package, whereas an aftermarket stereo with navigation can be had for less than $1000.

Another styling mod worth mentioning is a weighted shift knob. You may not realize it but the shift knobs in the Mazdaspeed 3's and 6's are already weighted from the factory. The benefit of a weighted shift knob is it's easier to shift due to the momentum of the weight once in motion. There are many styles and sizes of shift knobs on the market to choose from. One thing you should take into consideration is the material it's made from. Many shift knobs on the market are made from aluminum which is lighter than the factory weighted knob (which makes shifting harder). Also, the first time you grab a metal shift knob on a hot summer day, the experience will be "seared" in your memory for life. A proper shift knob should be leather wrapped, weighted, and compliment the interior styling of your Mazda.

As for lighting, I included a mod card in the last section for upgrading your exterior lights to LED's. The same thought applies to the interior bulbs. This is one mod, once completed, that will make you wonder how you ever put up with the ugly yellow light from the incandescent bulbs. Take advantage of 21st century lighting.

Lastly, I will offer up some useful options you should consider from your Mazda dealership. At the time of purchase you may have seen many accessories at the dealership. Some are handy, some provide extra utility, and others are just nice to have. Cargo mats and nets can help keep the hatch or trunk area tidy and looking sharp while providing protection. Floor mats allow a touch of style while giving protection to the very expensive carpet underneath them, and doorsill covers can add a touch of style to an otherwise drab part of a car.

SHIFT KNOB

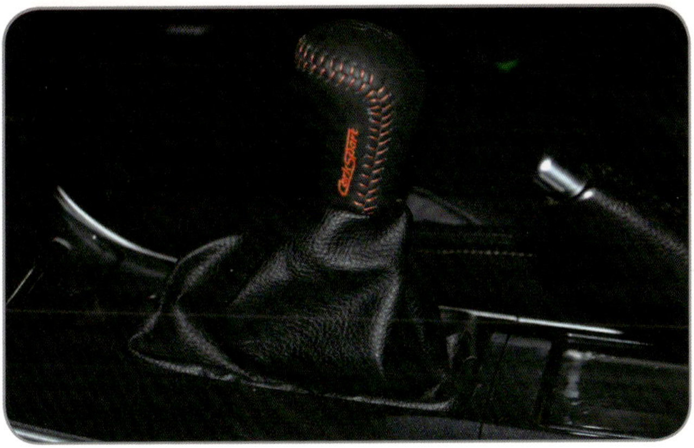

What it Replaces:

The factory shift knob with a heavier, more stylish knob

Why You Should Want It

- *Reduces shift times*
- *More stylish design picked by you*
- *More comfortable feel*
- *Easy to install*
- *Will not effect your warranty*

What You Should Look For

- *Leather wrapped*
- *Stylish design*
- *Heavier than factory weight*
- *Screws on, no set screws*
- *No exposed metal*

STEREO INSTALL KIT

What it Replaces:

The factory stereo and dash panel

Why You Should Want It

- *Improve sound quality*
- *Pick up the latest features like bluetooth, iPod integration, navigation, etc*
- *Allows for 2-DIN stereo for larger display*

What You Should Look For

- *Does not require cutting of factory wiring*
- *Optional connectors for steering wheel controls and BOSE audio for cars so equipped.*
- *OEM quality fit and finish*

INTERIOR LED KITS

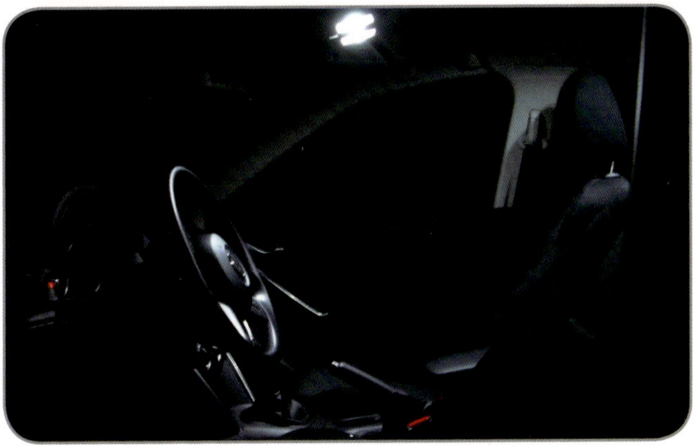

What it Replaces:

Replaces the incandescent interior lights in your Mazda with LED lighting

Why You Should Want It

- *Has a longer lifespan than other bulbs*
- *Has a lower power draw*
- *Resists shock and vibration better*
- *Generates less heat*
- *21st century lighting, vastly improved aesthetic*

What You Should Look For

- *A kit that will replace all the interior lights*
- *Highly respected vendor to ensure quality*
- *Installation instructions*
- *Retains interior dimming feature*

MAZDA GENUINE ACCESSORIES

Roger Beasley Mazda Central Austin, TX

There are plenty of options available at your local Mazda dealership. You can purchase protective items like floor mats, cargo nets, and cargo trays. You can upgrade your audio components and add tech package options to your Mazda, such as, in–dash 6–disc CD MP3 players, hand free devices, satellite radio, and navigation.

3.5 Packages

There are many thoughts and opinions on which parts to add first as you begin modifying your Mazda. In this section I am going to group the parts I have recommended on the mod cards (along with a few others) into what I will refer to as staged packages. This is what I believe are the best parts to consider first, second, and third on your way to achieving your goals. Many will stop at stage I, but some will want to continue to build out to stage III and beyond.

Just because I have put together packages, doesn't mean you have to buy one to buy the other. It also does not mean you need to have everything in stage I to get parts that are in stage II. This is simply a guide to get you from no mods, to fully built. I do believe most people will be satisfied upgrading parts in the order I present as staged upgrades.

For those just looking to do a few modifications, Stage I has the best bang for your buck mods which are the easiest to install. These are entry level parts and for those just starting out, a great way to get familiar with doing your own installs and adding aftermarket parts to your car. You can begin to determine how far you want to take your build by seeing how these mods change power and handling.

The stage II package provides the performance one might expect for a weekend racer. These are parts that will improve power and performance and even carry your car to victory at an autocross or track day. The stage II package has parts that will be more difficult to install and some could affect your warranty if you take your car into the dealership. This will be a good time to make a determination on what kind of risks you're willing to take. If your car is still under warranty you may want to wait until your factory warranty has expired.

Stage III packages are for the serious modifier. The stage III packages pretty much includes what you can "bolt-on" to your vehicle, before you start talking turbo kits. If you have made it to stage III, then you have likely become very experienced with both installations and modifying your car, and it is time to start talking "next level" mods which I will cover in the next chapter of this book.

For those people with lots of resources and experience, they might get all the Stage I-III parts at once or know enough to have developed a sense of which mods they like and which they don't.

I've not presented styling in stages here and by and large, these mods are subject to personal taste. However, don't neglect these parts. When I look back fondly at my car after parking, I don't see the sway bars, but rather admire how awesome my car looks on those new wheels!

The last point before hitting the mod cards is that I've split power mods for NA and Turbo cars. They respond too differently to mods to group them together.

NON TURBO STAGE I POWER PACKAGE

Short Ram Intake, Cat-Back Exhaust, Motor Mount Inserts, Short Shift Plate

This represents the first and best mods that you can add to your non-turbo Mazda to get some serious performance gains for a reasonable price. You can expect to spend in the realm of $850–$1000 for this package, the biggest expense is the exhaust system, so if you need to save for that, you can get the motor mount inserts, short shift plate and intake for $250–$350 and get started on your stage I upgrade.

NON TURBO STAGE II POWER PACKAGE

Lightweight Pulley, Header, Cold Air Box, Oil Catch Can, Rear Motor Mount

Some of these items are not found in the mod cards because they are not really bang–for–your–buck mods. Both the header and the lightweight crankshaft pulley are more advanced mods, but they can add power to your non–turbo Mazda. Some items are readily available and easier to install depending on the model of non–turbo Mazda you have. If you want to maintain a "Newbie" build, stick to the motor mount, oil catch can, and air box to get some solid improvements. You can expect to spend $1000–$1500 for these upgrades.

NON TURBO STAGE III POWER PACKAGE

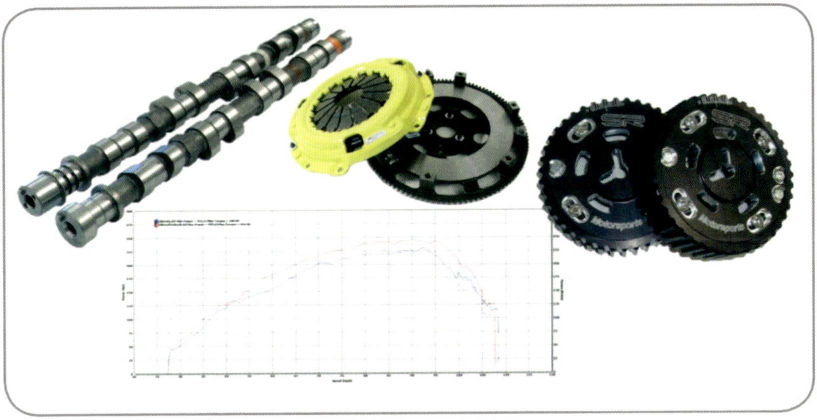

Camshaft, Lightened Flywheel, Cam Gears, and a Custom Tune

These are not on the mod cards due to the complexity of the installation and money that will be required. As with the stage II build, some items are readily available and easier to install, depending on the model of non–turbo Mazda you have. Some of these items may not be available for your non–turbo Mazda. I will talk about resources and places to look for parts in chapter 5. That will help guide you on your mission if you decide to you want to take your build to this level. These upgrades can cost anywhere from $1500–$2500.

TURBO STAGE I POWER PACKAGE

Rear Motor Mount Inserts, Short Ram Intake with Turbo Inlet Pipe, Cat Back, Exhaust System, Short Shift Plate, Oil Catch Can

This represents the first level and best mods that you can add to your turbocharged Mazda to get some serious performance gains for a reasonable price. The biggest difference between this and the non–turbo package is the addition of the turbo inlet pipe and the oil catch can. In turbo Mazdas, blowback from the crankcase is a common issue that you can address with the oil catch can (OCC). You can expect to spend in the realm of $1000–$1500 for this group. You can get the motor mount inserts, short shift plate, intake, and OCC for $450–$600 and get a start on your stage I project if you need to save for your exhaust system.

TURBO STAGE II POWER PACKAGE

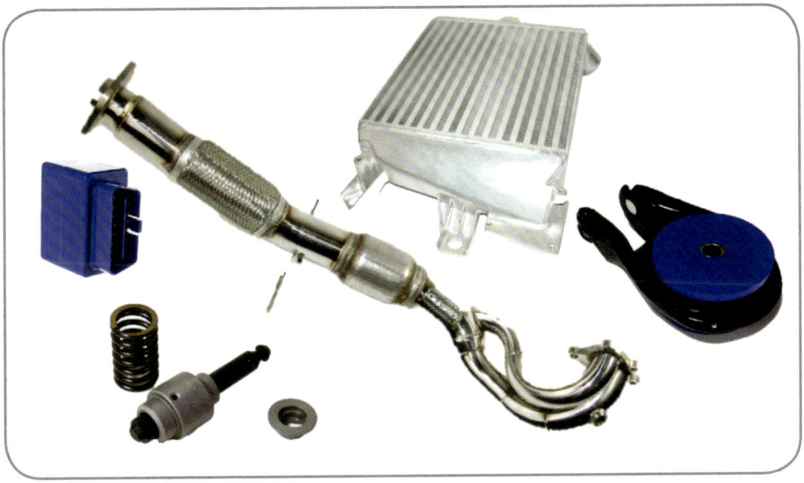

Catted Downpipe, TMIC, Blue Tooth OBDII Monitoring, Fuel Pump Internals, Rear Motor Mount Replacement

There is a school of thought out there that you need to add a tuning device at the same time you upgrade your fuel pump internals, however my experience shows that upgrading just your fuel pump internals at this time and adding a blue tooth monitoring device will save you money and provide the piece of mind to maintain engine health and fuel trims. If you plan to continue to upgrade beyond the scope mentioned in this chapter of the book, you may want to consider skipping the Top Mount Intercooler and get a Front Mount Intercooler instead. Expect to spend $1500–$2000 for these mods.

TURBO STAGE III POWER PACKAGE

Tuning Device, BOV or BPV, Engine Mounts, Big MAF Intake Kit

This would be the final stage of "bolt–on" modifications before going big turbo. You will certainly want to get a tuning device and may even want to consider a custom tune at this point to get the most performance possible from all the parts you've installed. At stage III, keeping an eye on everything is very important; having a proper tune can mean the difference between fun and failure. Expect to spend $1500–2000 for these mods.

HANDLING STAGE I

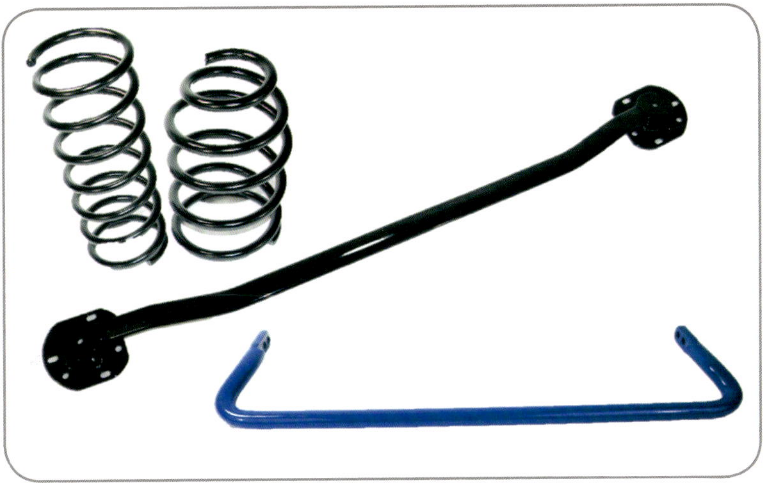

Lowering Springs, Front Strut Bar, Rear Sway Bar

No matter what you drive, adding springs, a front strut bar, and a rear sway bar to your Mazda will provide immediate and noticeable improvements in handling. For this kit, you are looking at $500–$700. If you are using your car as your daily commuter, this may be all you need for a perfect balance of performance and practicality. Wheels and tires, also, are great entry level handling mods, but because most add them as a styling component, I have those listed under styling mods and did not include them here.

HANDLING STAGE II

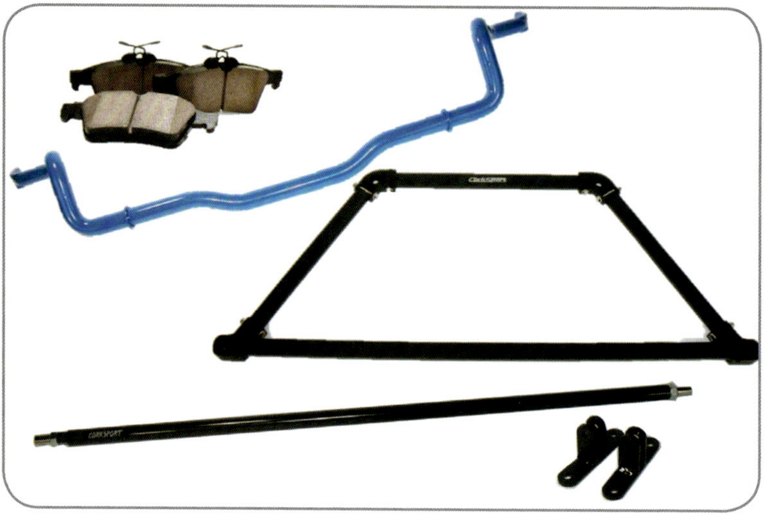

Brake Pads, Front Sway Bar, Rear Stress Bar, Chassis Bracing

The next level of suspension upgrades includes a front sway bar rear strut bar or rear chassis brace and additional sub–frame bracing. You could consider a mid–chassis brace, front arm bar, or, essentially, any other part that will tie your suspension or chassis points together. In addition to added bracing, I would also recommend upgrading your brake pads at this time for better stopping power. Expect to spend $500–$700 for these additions.

HANDLING STAGE III

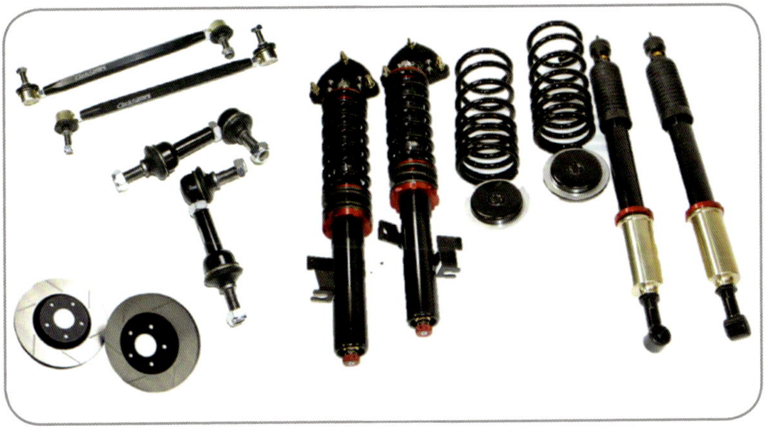

Front and Rear Endlinks, Coil-overs, Brake Rotors

Finally, for those serious about handing; adjustable coil-overs. I would recommend adding adjustable front and rear sway bar end links and upgraded brake rotors. Notice the word adjustable is used a lot here, that is because you can fine tune your suspension components to your exact driving style. Racing can come down to a few seconds and suspension tuning can be the difference. There are plenty of people out there that will get coil-overs just to slam their ride, but in reality, adjustable coil-overs are the best option for a car driven on the street and track. If camber plates aren't included in your coil-over set–up look at adding that to the set–up for even more adjustability. Full setups like this can range from $1500 all the way to $5000 or more.

CHAPTER FOUR

TAKING YOUR MODIFICATIONS TO THE NEXT LEVEL

I have covered what I think are the most common and beneficial performance modifications. If you are interested in going to the next level, I'll now cover more advanced upgrades. It is important to keep in mind, that many of these items require other upgrades to support them. Installing a turbo kit without upgrading your fuel pump internals, adding a big turbo intake, and a tuning solution can be extremely dangerous and could result in engine failure, or as it is commonly referred to on the forums, "ZOOM ZOOM BOOM". Before moving along with the advanced mods, I'd like to share with you a customer's story to tie this all together.

Evan McIntire used to stop by and visit me regularly at my shop (he lived just down the road at the time) and keep me company in the showroom while getting some mods added to his car. Evan didn't exactly follow the staged upgrade path I presented earlier, but he did work closely with me and blazed his own trail in a smart way.

Evan has since moved to Honolulu, HI to pursue Biology at the University of Hawaii at Manoa.

Shortly after he got his drivers license, Evan got a hand-me-down 1994 Honda Accord. When the thrill of just being a new driver wore off, he knew he wanted to upgrade and couldn't get the thought of a Mazda 3 out of his head.

New to driving, Evan didn't know how to drive a manual so he never considered a Mazdaspeed 3, but his brother kept telling him the speed 3 was the way to go. The thought of a 300 WHP speed3 is a strong incentive to learn, so he learned to drive a manual and got his 2007 Mazdaspeed 3.

He used to enjoy sitting around and watching his older brothers work on various cars with their friends until late into the night. He would always ask how he could help and how things worked. His Mazdaspeed 3 was his first real experience modifying his own vehicle.

Evan started modifying his car with bang for the buck power mods, a Short Ram Intake and a Turbo-Back Exhaust. Urethane exhaust hangers were added to keep the larger exhaust in place. An HKS SSQV Blow Off Valve and a COBB Tuning AccessPort (to remap his fueling) were also added.

Next were handling modifications. He added Lowering Springs to give his Speed a more aggressive look and improved handling. Rear Motor Mounts were added to help keep the engine firmly in place and offset all the newly added power. With improved handling and power came the need for improved braking, so Evan added stainless steel brake lines. He also smartly added some protection for the undercarriage of his vehicle now that it was lowered by adding a skid tray.

Next he focused on some transmission area upgrades with a Short Shift Plate and Shifter Bushings. He then focused on more aesthetic items by adding headlight eyelids, a front license plate relocation kit, vinyl from his favorite Mazda Performance Company, upgraded interior lights, and fog lights, as well as added an LED underhood lighting kit, which is especially useful for seeing underhood at night.

A lot of people are happy to stop at this point, but Evan wanted more, so he took his build to the next level and added an aftermarket crankshaft pulley, a Front Mount Intercooler Kit, EBD GD Series Brake Rotors, Red Stuff Ceramic Brake Pads, and a Full Replacement Short Shifter.

Watching Evans car come together the way it has and the smart decisions he has made along the way to balance power, handling, braking, and styling is exactly how I like to see a build up occur. He has put a lot of money into his build, but has done it gradually over time

And as Evan put it, "I love how no matter what is going on in my life, I can always get in my car and go for a drive, day or night, and during that drive nothing else matters but the road that lies ahead."

Evan is a great example of how far a typical enthusiast goes with his performance mods. In this chapter, I will hit on most of what he has added to his car, not already on the mod cards and talk about why they are smart modifications to make.

I was happy to provide Evan with the resources needed to meet his goals and hope this book will be that resource for you. I still miss his company, but I don't blame him for trading the dreary rain of the northwest for sunshine and paradise.

4.1 Turbo Kits

If you want to boost your Mazdas power above and beyond bolt-on's you'll want to add a turbo or upgrade to a bigger one. I'll talk about both upgrading a car already turbocharged as well as one that has no turbo on it from the factory.

Upgrading the Turbo on an already Turbo-Charged Mazda

For vehicles like the Mazdaspeed 3 and 6, a turbocharger upgrade is commonly referred to as a Big Turbo Kit (BT). The amount of air moving through your engine is what ultimately determines the total power output. By adding a larger turbo to your vehicle, you will be able to increase the air moving through your engine to achieve additional power gains.

There are several factors to look at when going down this path. First is how much power do you want to add? Second, how much do you want to spend? Your power goals will dictate how much you will need to spend. Once you've decided you want a larger turbo, your costs go up exponentially, so I'll mention a few things to consider.

Option A:
Bolt-in Upgrade kit for factory exhaust manifold

First, more power means a larger turbo. As you go larger with the turbo you shift max power to a higher RPM. If the factory turbo starts adding power at 2500rpms, a slightly larger turbo might start adding power at, 2700rpms. But if you go with a very large turbo your power might start coming on at 4000 or 5000rpms. This is not very practical for a daily driver. It results in a car that is basically gutless most of the time then power comes on so fast that tires just spin. It's good for showing off, but not very practical.

The second thing to think about is all the other parts this turbo upgrade will require. There are bolt in kits available that use the factory exhaust and intake manifolds that will give a nice bump in power, perhaps 30% over stock. This sort of BT gives you easier installation combined with a good power increase and you are still in the power range that the factory motor might be able to withstand. This manner of turbo upgrade keeps you in the $2000-$5000 range.

Option B: Custom Kit

If you decide you want to go with a much larger turbo you need to really sit down and plan out your costs. A larger turbo will not directly bolt in to the factory location. Even if the turbo only costs you $2000 you will need to put aside quite a bit of money for the extras. You also need to remember that when you add a larger turbo, your motor will wear much more quickly and can be more prone to failure. The factory pistons and rods are not designed for this sort of power. Upgrading your engine internals is a good idea. This can cost $3000 on up depending on the parts and labor involved. You can easily drop $10,000 on a custom large turbo setup. Also, think about down time; your car will not be running while you upgrade all these parts. Even with the best planning you can expect to be without your car for a week, maybe a month or more! You might also want to own a separate daily driver if you are going nuts with a custom setup, as they require lots of tweaking and there are always various issues that arise.

Most of the larger turbochargers use an external wastegate (remember question #8 in chapter 2). External wastegates are outside the turbo (hence the name), and are normally fitted on the exhaust manifold. This requires custom exhaust work to route the wastegate to the downpipe or vent it towards the ground under the car.

Parts List for

Turbocharger Upgrade, $2,000-$10,000

- *Turbocharger*
- *Custom dyno tune*
- *Big mouth Intake with Turbo Inlet Pipe ***CUSTOM****
- *Upgraded clutch and pressure plate*
- *Slush fund, $500*

OPTIONAL (but will limit how far you can go if you don't)

- *Exhaust Manifold ***CUSTOM****
- *Exhaust plumbing or new downpipe ***CUSTOM****
- *Intake Manifold ***CUSTOM****
- *External Wastegate*
- *Fuel Injectors*

The last piece for a turbo upgrade is getting your Mazda tuned to take advantage of the new found airflow volume. This can be handled by taking your Mazda to a tuner near you or contacting one of the many e-tuners available on the market who have experience with upgraded turbos.

Upgrading a non-turbo Mazda to a turbo Mazda

There is a distinct power advantage to a turbocharged engine over a naturally aspirated (NA) engine of the same size. Naturally aspirated engines suck in air whereas a turbocharger is basically an exhaust-driven pump that crams air into the engine.

Adding a turbo to a non-turbo Mazda will definitely get you more power, well past what adding even Stage III parts will get you, however there is a large cost associated with this undertaking.

Option A: Off the Shelf Turbo Kit

The cheapest path for an NA car looking to go turbo is to trade in your non-turbo Mazda for a turbo Mazda (which might be cheaper in the long run). Assuming you won't do this, the next best option is to find a turbo kit which is made for your Mazda. A complete kit would include all the pieces you'll need including a turbo, an exhaust manifold, intercooler, turbo inlet pipe, downpipe, blow off valve, piping, and all the needed hardware and hoses. This can run you anywhere from $2,500-$5,000 for the kit. There are also some hidden costs that most people don't discover until they are neck deep in their project. Even the best mechanic will not likely complete this in a weekend. For most there car will be down for several weeks to months. You also need to budget extra money for the unknown issues you will run into.

Option B: Custom Turbo Kit

The other option is to build a custom kit yourself. The toughest part of the DIY option is locating an exhaust manifold to bolt up to the turbocharger. Popular Mazda models may have manifolds available whereas other models will not and require you to make a custom manifold from scratch. This can be extremely time consuming and expensive.

Parts List to add

$7,000-$10,000

Turbocharger NA car, no kit available

- *Turbocharger*
- *Wastegate*
- *Intercooler*
- *Intercooler Piping, silicone connectors, clamps ***CUSTOM****
- *Piggy back or stand alone ECU with tune*
- *Upgraded Fuel Pump*
- *Fuel Injectors*
- *Intake with Turbo Inlet Pipe ***CUSTOM****
- *Blow Off Valve*
- *Exhaust Manifold ***CUSTOM****
- *Turbo-back exhaust ***CUSTOM****
- *Upgraded clutch and pressure plate*
- *Slush fund, $1000*

While the list is not comprehensive and there is a wide range of money you could spend on the various components, the shopping list on the preceding page will give you an idea what you are up against. Some components could be sourced used (see inherit someone else's problem) or sourced cheaper (like a turbo back exhaust at your local muffler shop). You get what you pay for, so I tried to stay somewhere in the middle of the cost versus quality spectrum. ***CUSTOM*** means these parts are very often hard to find, and if you have to make it specifically for your car, costs will increase substantially. These lists do not include labor!!!

Turbochargers are easy to find as are intercoolers and the piping needed to hook everything up. You should have access to a welder. Tuning options for the newer model cars can include a re-flash of your ECU, a piggy back ECU, or the last option; replace your ECU with a standalone EFI system. The easiest route is to re-flash. This means no wiring or splicing like you would need in a piggy back or a standalone ECU. If you have to go that way, the build can get significantly more complex.

It will likely take you 15 or more hours to install a turbo kit (at a minimum), and you will want access to a lift in order to perform the install. If the install doesn't go right your car may be down for weeks.

Don't Blow Up Your Engine!

You should carefully consider the increase in pressure inside the cylinders that adding or upgrading a turbo system can create. A safe increase for most 4 cylinder Mazda engines is around 50 horsepower, and 5-8psi of boost. If you push the limits too far you can join the Zoom-Zoom-Boom club. You can easily increase the factory power by 50%. This will result in additional wear on internal parts such as rings and seals. Expect to rebuild your engine sooner than you would if you had left the engine in factory form.

An upgraded fuel pump and/or fuel injectors could be required. This is critical since the original fuel delivery system was not intended to support that much extra power.

What you should really take away from this section is this, if you intend to do this upgrade then plan, plan, plan. A small increase in power is the way to go if you are on a budget. If you want to go big you need to expect added costs and added downtime.

4.2 Cooling

Cars create heat; they come equipped with radiators, coolers and intercoolers to help get rid of some of it. Stock components are made for typical conditions and stock parts; once your power levels rise, those stock components can become overwhelmed. Even longer duration spirited drives can over-tax some of these stock components. Fortunately, many of these items can be easily upgraded.

Intercoolers for Turbocharged Mazdas

Turbochargers substantially increase engine air temperature as they compress the air. This hot air can easily lead to engine damage from early detonation or "ping". The compressed air needs to be cooled before it enters the engine, this cooling comes from an intercooler whose basic function is to cool air that is heated up when compressed by the turbocharger.

The intercooler operates much like a radiator but cools engine air instead of coolant. A radiator is an air-to-liquid cooler, whereas an intercooler is an air-to-air cooler. The compressed hot air from the turbo enters the intercooler and transfers some of its heat to the internal fins of the intercooler while passing through and air from the outside cools the external fins on the intercooler.

There are several different types of intercooler upgrades to choose from. With the DISI engine used in the Mazdaspeed 3, Mazdaspeed 6 and CX-7, the most common options on the market are a top mount intercooler upgrade (TMIC) or a front mount intercooler replacement. See Mod Card for TMIC.

Front Mount Intercooler

A front mount intercooler offers a primary advantage over the top mount intercooler, because you are moving the intercooler from the top of the hot engine to outside of the engine bay where the fresh cool air is.

By moving the intercooler out of the hot engine bay and to the front of the car, the intercooler gets much cooler air and is able to do its job of cooling the charge air much more effectively.

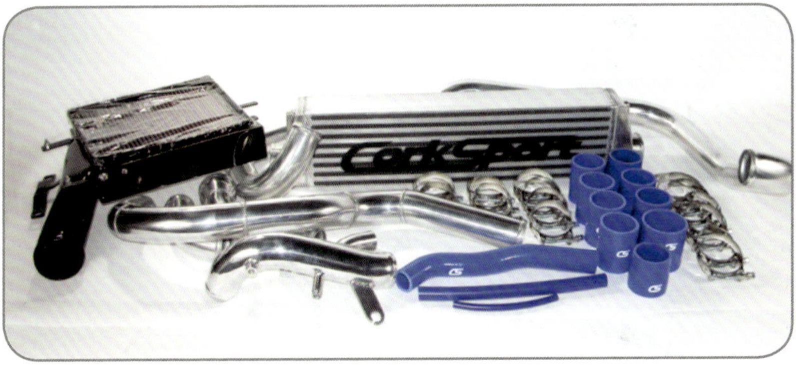

Radiator

"When do I need to look at upgrading my radiator?" The stock radiators used in Mazda's are designed for stock power and normal driving habits. When you start making significant upgrades to your Mazda (beyond stage III), your radiator may need to be upgraded to keep up. Bar and plate radiators with folded aluminum end tanks are popular options for these upgrades.

4.3 Clutch and Flywheel (Manual Transmission Only)

Once you start making a lot more power, one of the first things that will need some attention is your clutch assembly. The clutch has to be able to hold all the power from your engine and transfer it into the transmission. Upgrading your clutch will allow it to operate more efficiently, resulting in quicker acceleration and improved pedal responsiveness.

- **Upgrading your Clutch Disc**. The disc has friction material on both sides. The friction material can come in a variety of different substances just like brake pads: organic, ceramic and carbon Kevlar to name a few.

- **Upgrading your Flywheel**: At idle, the mass of the flywheel, actually, helps smooth out throttle inconsistencies since it takes effort from the engine to speed it up or slow it down. Aftermarket aluminum flywheels (referred to as lightened flywheels) will improve throttle response right away because the engine revs quicker and, thus, accelerates more readily. The one drawback is that as you release the clutch, the idle drops faster, making it easier to stall your Mazda. Little throttle inputs are noticed more readily.

Popular Pressure Plate and Lightened Flywheel

- **Upgrading your Pressure Plate**: The clutch disc is sandwiched between the flywheel and the pressure plate. The pressure plate is generally rated by how much torque it will hold. You should pick a pressure plate that will hold the amount of torque you plan to make, but not too much beyond that because pedal pressure increases as the ability to hold torque increases.

You don't want to be in rush hour traffic with a clutch that is next to impossible to push in. Always try to get a clutch that is rated slightly above to your engine's capabilities. This is another case where the biggest and baddest is not always best.

4.4 Engine Management and Custom Tunes

Tuning Q&A with Tuner Rob Finkle: Username rfinkle2 on Mazdaspeedforums.org

Having had many factory turbocharged cars before my Mazdaspeed3, I believe that a custom tune is the best modification that can be done from a performance per dollar standpoint.

What is custom tuning/what does ECU remapping do?

Ecu remapping, a/k/a tuning, are terms used to describe the process of making changes to the instructions that a car's brain (Engine Control Unit) follows. Remapping / tuning alters the variables that control the way the car behaves under a variety of conditions, including startup, idle, part throttle and WOT (wide open throttle). To "remap" a car, special software is used to communicate and alter the directions the ecu follows. The most dominant and reliable software for the Mazdaspeed platform is made by either COBB or VersaTune.

Why would you want to tune your vehicle?

When Mazda engineers tuned the factory vehicles, they were concerned with a broader set of goals than the average enthusiast. The engineers at Mazda know that each Mazdaspeed will be used in a variety of weather conditions, under a variety of user demands, and must be repaired under warranty if parts prematurely fail.

Mazda is also heavily concerned with emissions and gas mileage and balances the stock map by making compromises to meet these goals with the result sometimes coming with less power than the engine is capable of.

The enthusiast usually sees performance as the most important goal, while weighting reliability, gas mileage and emissions lower in terms of importance. If you want to maximize the performance, especially of a factory turbocharged vehicle, many "nannies" or electronic performance limitations need to be removed. Overly conservative measures in a stock map can be reprogrammed and can result in substantial gains in horsepower and torque.

Who should tune their vehicle?

This is an interesting question, I believe both reliability and performance can be increased with a well thought out tune. In my opinion, if budget allows, anyone who owns a Mazdaspeed can benefit from a custom tune.

What gains can be expected?

While every car is different across the Mazdaspeed platform, in general, if the 93 octane gasoline in your area is of decent quality, a stock Mazdaspeed can pick up 5-10% more power just from a tune.

On an OTS (off the shelf) map, cars with an upgraded fuel pump and common modifications usually make 10-15% more power at the wheels. Custom tuned modified cars usually get 5-10% more than that.

A custom tune for e85 octane gas, can net yet another 10%. For you math majors, a Mazdaspeed with common bolt-ons and a custom E85 tune can get around 30% more power than stock.

What is the benefit to a custom tune vs. a chip or power programmer?

Much like Mazda engineers, creators of maps included on custom programmers such as Hypertech etc. are responsible for making maps that will run well in all weather conditions, at all altitudes and with many different combinations of bolt-on modifications. A custom tune can take full advantage of your specific environment and every modification on your car. Custom tuners can tune the car to perform and feel just about any way the owner wants. Boost onset, boost level, boost level per gear, engine idle speed and pedal feel all can be customized to your preference.

Does it have an effect on your warranty? Are there hidden costs and drawbacks?

These are a fuzzy questions in terms of the relationship to reliability and warranty. As discussed above, I'd argue that a well tuned Mazdaspeed will be more reliable than one that is running the "one size fits all" Mazda map, due to the ability of the tuner to customize a map for driving style, altitude etc.

As far as the Mazda warranty is concerned, the tune is not usually a point of contention with Mazda, as most warranty rejection is due to installed hard parts (which can be uninstalled before dealer visits).

In short, after returning the stock ROM program onto the ecu, experience suggests that the Mazda techs can only tell the ecu has been reset if there are a limited number of miles put on the car since the last reflash.

If you are a Mazdaspeed enthusiast in search for more horsepower for your Mazdaspeed, the performance per dollar spent on a custom tune is hard to beat by any individual hard part.

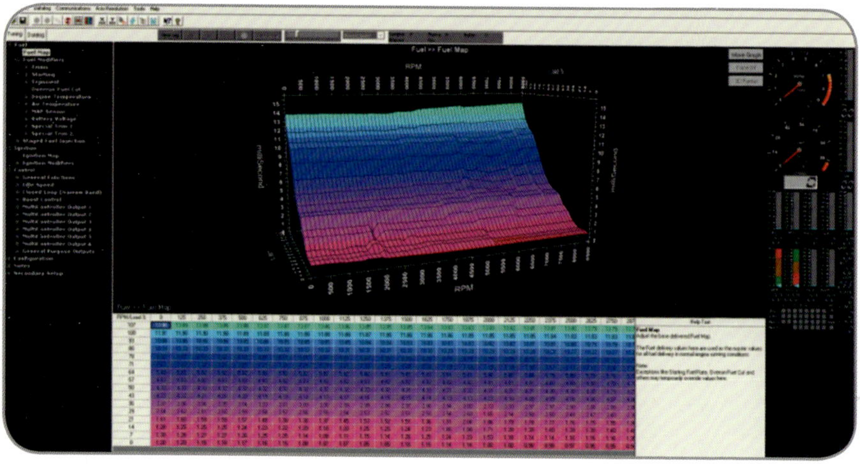

Screenshot From a Popular ECU Reprogramming Software

I'm forever on mazdaspeedforums.org, so if you have questions, feel free to send me a private message. If I don't know the answer, I'll do my best to help you find it.

4.5 Nitrous

Nitrous

Earlier I advised against nitrous (Nitrous Oxide). However, now that you are more familiar with your car and the modification options, perhaps you are left wanting more power than these 'bolt-ons' can give you. A properly installed nitrous system can give your Mazda a big power boost. For instance, to pick up 100 horsepower on a naturally aspirated car you might have to spend $1000's whereas you could get that power for a few hundred dollars with nitrous.

Nitrous works by providing extra oxygen for combustion. Nitrous systems need extra fuel to burn. This comes via an additional fuel injector or by making your original Mazda injectors flow more fuel. The resulting power can be quite amazing, but can put lots of extra stress on your engine unless setup and maintained perfectly.

You'll have to run different spark plugs, and since most of the time the nitrous will be off, the 'colder' plugs needed for nitrous applications can foul which leads to misfire and eventually a disabled car. Furthermore, you'll need to run the best gas you can get.

Another big catch, nitrous costs about $5 for every 10 seconds it's used. One impulsive pedal to the floor to zoom-zoom ahead of that jerk that cut you off costs more than a Venti Iced Mocha with an extra shot or three cans of your favorite energy drink.

Some people think of nitrous as cheating. Frankly, that's probably why I have not "squeezed" myself but it you want to drag race or just have some power on tap for 'emergencies' you can give nitrous a closer look.

4.6 Suspension

There is a great deal more to your Mazda's suspension than shocks and springs. Once you have upgraded your suspension with strut tower bars, sway bars, and coilovers; as discussed on the mod cards, you may find you still want to continue to improve the handling of your Mazda. Bushings, underbody braces, end links, and other parts can be great for fine tuning the handling for a further customized driving experience.

Bushings

The factory uses soft rubber bushings in suspension components to minimize vibration and road noise. Upgrading to aftermarket bushings can improve cornering and driver feedback. Most are made of polyurethane and can come in a variety of hardness characterized by a term called durometer, the higher the durometer the stiffer the compound. Some factory bushings can even be replaced by spherical bearings that are solid.

Bracing

There are a variety of bracing options that connect your sub-frame, open areas of the chassis, or stiffen your control arms. Some of these can actually save you money over time by lengthening the life of bushings. When you see a race car there are two reasons it has a cage: safety is one, stiffening the chassis is another.

2011 Mazda2 B-Spec Prepared Race Car

Unlike a roll cage, braces are an easy alternative that doesn't require permanent modifications to your car. Look for items like chassis braces, lower arm bars, and 4-bar bracing kits.

For the ultimate suspension tune, you can get pieces that will allow you to dial in the handling to your preference. You can purchase camber plates, camber arms and toe arms that will allow you to adjust your car for track days, then restore for daily driving so you are not wearing your tires so quickly.

Adjustable end links allow fine tuning of your sway bar for that perfect setup

4.7 Most Common Mods I Skipped and Why

You've now looked at a long list of entry level mods as well as things to consider next if you want to take your car to maximum performance. You should have a better understanding of the basic mods and hopefully a bit more of an idea of the direction you'd like to go with your car.

Perhaps you are thinking about some parts that I didn't mention at all in this book. My feeling is that if you know about those mods or are interested in performing them then you are likely already past the scope of this book. I've tried to keep things simple here and give a list of the most common, entry level mods you can perform on your Mazda.

Some examples of the types of mods I've skipped would be intake/ exhaust manifolds, meth injection, camshafts, engine internals, superchargers, and other suspension mods like air ride. The majority of people aren't building full tube chassis race cars with 4 rotor race engines and a professionally tuned suspension so there isn't a lot of need to write about those items here. I kept this book short, sweet, and to the point. The more advanced modifications are for another time and place.

CHAPTER FIVE

LEARNING MORE

So now you know where to start, how to start, what the parts are, and things to look for. Where do you go to learn more? In the pages to follow I'll talk about where to get the parts previously discussed and how to be sure you're buying from a trusted source. I'll also go over some things to look out for when purchasing.

When we're done talking about what to look out for, I'll give you some info on places to go for more guidance as well as places to look if you're interested in getting involved in your local Mazda community. No matter where you live there is likely a Mazda community nearby. I'll give you some ways to get out there and enjoy your Mazda while surrounding yourself with people that will likely be able to help you learn more about your car. I'll discuss Car Clubs and how they may benefit you, as well as the different types of organized racing such as SCCA and Auto-X (autocross).

5.1 Where to Get Parts and Things to Look Out for When Purchasing

Where to get parts

There are lots of places to source aftermarket parts for your Mazda. It's possible to go to your local auto parts store and purchase some of the most popular parts made by some of the most respected names in the business. If you happen to live in a large city, you may even have a performance shop in your town that can provide advice, sell parts and even install them for you. In a very few, select cities you'll even find a shop that specializes in Mazda performance or at least compact cars or imports. Of course, perhaps the most popular way to source parts these days is to hit the internet and type "Mazda performance" in Google.

It's hard to tell which companies on the internet are legit, and have the wherewithal to stand behind their products. If you mod long enough, you will run into a problem with aftermarket parts. I'm going to save you some time and frustration by listing some of the best companies I know that specialize in Mazda performance parts and accessories. I'm not listing companies that are primarily resellers below (like Amazon and eBay). Once you decide on a part that meets your needs you can probably find many different places to buy it. The companies below are the most long standing and well respected names in the business* (alphabetical order)

COMPANIES SPECIALIZING IN MAZDA PERFORMANCE PARTS AND ACCESSORIES

COMPANY	LOCATION	SERVICES
AutoExe A New Driving Sensation	*AutoExe Japan*	*Full line of parts for all Mazda cars, expensive though.*
CONTACT *http://www.autoexe.co.jp/english*		
AWR	*AWR California, USA*	*Full line of parts for Mazda cars*
CONTACT *www.awrracing.com • 760-598-9782*		

COMPANIES SPECIALIZING IN MAZDA PERFORMANCE PARTS AND ACCESSORIES

COMPANY	LOCATION	SERVICES
CorkSport Mazda Performance	*CorkSport Washington, USA*	*Full line of parts for Mazda cars*
CONTACT www.corksport.com • 360-260-2675		
Flyin' Miata	*Flyin' Miata Colorado, USA*	*Historically, a Miata specialist*
CONTACT *www.flyinmiata.com • 800-359-6957*		
KNIGHT SPORTS	*Knight Sports Japan*	*Historically, a rotary specialist*
CONTACT *http://www.knightsports.co.jp*		
MAZDASPEED MOTORSPORTS DEVELOPMENT	*Mazda Motorsports California, USA*	*Supports Mazda racing in the USA and beyond. Racers only!*
CONTACT www.mazdamotorsports.com • 800-435-2508		
MAZDATRIX	*Mazdatrix California, USA*	*A rotary specialist*
CONTACT www.mazdatrix.com • 562-426-7960		
RACING BEAT	*Racing Beat California, USA*	*Historically, a rotary specialist*
CONTACT www.racingbeat.com • 714-779-8677		

These companies specialize in only Mazda Performance and have been around more than 10 years.

Why purchase from these companies when there are other options? First, Mazda is a relatively small brand. It is a niche company. As such, you'll find the depth and breadth of aftermarket parts lacking compared to other brands like Honda and Subaru. Even popular models like the Ford Mustang and Chevy Corvette have larger aftermarkets than the entire Mazda Brand. So when you buy from these companies (rather than a product line manufacturer like K&N, Magnaflow, or Eibach) you help grow the Mazda aftermarket.

What to look for

Here are a few questions or points of consideration when selecting a company to purchase aftermarket parts from.

1. Are they easy to reach? Many companies are very slow to respond to emails and often don't answer their phones consistently.

2. Do they offer technical support? If service before the sale is not great, you are unlikely to get great service after the sale!

3. Do they offer a warranty? What is it?

4. Are they active in the Mazda community? Check out the next section for more on this.

5. Do they provide quality installation instructions for the Do-it-yourselfer? Can you view them online before you purchase?

6. Do they ask for and listen to your input in product development?

7. Are they well respected and well-reviewed?

5.2 Resources and Community

With the internet it's easy to find information on your Mazda from local clubs, internet forums, and companies. If you are looking for more information beyond what's presented in this book be sure to check out some or all of the sources below.

Local car clubs have been around almost as long as the automobile. They are groups of enthusiasts who are in close geographical location to each other, that share the same passion for their Mazdas. Most have activities each month such as meetings, drives to a destination, car shows, and some even go racing. Clubs can give you a hands-on approach to learning about your car with other club members and can provide support by working on your Mazda during an install day.

Local Owners' Clubs

Below is a list of just a few clubs out there. Most have webpages and forums. Find out what they are doing and get to know them.

- **NATOR** Nator is more of a fraternity of car enthusiasts with groups scattered all over the US. To find your local NATOR group check out the Mazdaspeed Forums (more on that soon).

- **MazdasNW** MazdasNW is a group of enthusiasts who welcome any Mazda. Most of their members are located in Washington, Oregon, and Idaho.

- **MOCC** The Mazda Owner Club of California is one of the biggest clubs out there. They have been hosting the largest Rx7 meet, Sevenstock, for many years.

- **SFLA** The South Florida Mazda Crew is a great group of enthusiast. I have had the pleasure of attending their events in the past. They welcome any Mazda owner.

- **Miata Clubs** There are over 100 of them. Do a Google search for Miata club, include your area, and you will find one. I could fill up a page adding them all in.

- **Texas Speed Demons** https://www.facebook.com/Texas.SpeedDemons A group of Texans dedicated to Mazda and each other.

Internet Forums

Internet forums are online communities where it is easy to exchange technical information, ideas, and talk to other people who have the same car you do. It is also a great place to make friends and find out about events near you. I have met some great people on the forums who I will know for years to come. Most of the forums have technical sections or FAQs to help you find answers to common questions. Consider reading this book again before diving off into the forums so you can get the most out of your time spent there. The glossary will help you out a lot.

Most forums cater to one specific vehicle or group of vehicles. The list below is a good place to start but is by no means a complete list.

- **Mazda3revolution.com** Geared towards the 2010+ Mazda 3s and Mazdaspeed 3s.

- **Mazda3forums.com** Covers Mazda 3s, all years.

- **Mazdaspeedforums.org** The Mazdaspeed forums focuses on the Mazdaspeed 3 and Mazdaspeed 6. There is a great amount of information there, just make sure you use search first before asking questions as the natives can be unkind to "noobs".

- **Mazdas247.com** This forum has sections for every model of Mazda. It is a strong forum for the 99-03 Protégés and Mazda 2s.

- **Clubroadster.net** Have a Miata MX5? Give these guys a look. There is a good amount of tech along with some crazy ideas in there.

- **Rx7club.com** This is the home of the Rx7s and other rotaries. It has been around forever so there is a good chance someone has already asked your question before.

- **Rx8club.com** Home of the Rx8s and by far the most popular of the forums for the most recent rotary.

5.3 Common Check Engine Lights

All cars eventually will have a check engine light flashing at you. Many times a check engine light is not the catastrophe you might fear it to be. In some cases, a check engine light (or CEL) can be the result of something small, perhaps a sensor running just outside of its' accepted threshold, or an oxygen sensor that has yet to warm up and is not reading properly.

In order to pull codes simply plug in your OBDII reader or adapter. OBDII readers will generally have a scan button. After scanning make sure to write down all codes before using the clear feature to clear them. The codes will continue to come back until something is fixed. For OBDII adapters that use free applications like Torque you can simply start the program and you will go through the steps of scanning and clearing. The on screen instructions will help you with any application or device.

Many auto parts stores will gladly help pull your code, often without charge.

Common Engine Codes

- **P0139 Oxygen Sensor Circuit Slow Response (Bank1, Sensor2)** – You will commonly see this if your second (rear) O2 sensor is going bad or because of a lack of a catalytic converter.

- **P0171 System too Lean (Bank 1)** This is a common issue when you have a vacuum leak in the intake system. It is important to address this issue because your car is not running correctly. Many times it's because clamps are not tightened properly or vacuum lines were not put back on after a part swap.

- **P0172 System too Rich (Bank 1)** – Like P0171 but when you have a vacuum leak post turbo.

- **P0300 Random/Multiple Cylinder Misfire Detected** – This can be caused by a variety of reasons from bad gas to worn out spark plugs. When modifying a car this can occur on cars that need to be tuned because of mix and match parts.

- **P0301(0302, 0303, 0304) Cylinder 1(2,3,4,) Misfire Detected** – Similar to the P0300 but cylinder specific. This is generally from tuning issues or worn replaceable engine parts such as spark plugs, air filters or bad fuel.

- **P0420 Catalyst System Efficiency Below Threshold (Bank 1)** – Commonly caused by old, worn out or missing catalytic converter.

- **P0421 Warm Up Catalyst Efficiency Below Threshold (Bank 1)** – When your catalytic converter starts to fail the pieces break up inside causing slow exhaust movement.

- **P0442 Evaporative Emission Control System Leak Detected (small leak)** – Gas cap is not on all the way or there is a small vacuum leak in the emissions control system.

- **P2096 Post Catalytic Fuel Trim Bank 1 System Too Lean** – This shows up when you have a large difference between what the front and rear o2 sensors read. This is generally when you take the o2 sensor too far out of the exhaust.

5.4 A Brief Mazda History

The Toyo Cork Kogyo corporation was founded by a group of Hiroshima investors in 1920 who shortly thereafter put Jujiro Matsuda in charge. Originally tasked with manufacture of a cork substitute made from Abemake trees, this endeavor was eventually to become known as Mazda. As is often the case with very old companies, their name and primary product changed many times.

In 1930 their first wheeled vehicle was produced, a motorcycle. Out of this project a tri-cycle (cooler than it sounds) was born which they named Mazda.

Somehow spared from the atomic bomb dropped on Hiroshima to end WWII, the company's vehicles continued to grow in size until eventually they had four wheels. 1960 is the year that Mazda, as you or I think of it, got it's legs with the introduction of the R360 two-seat passenger coupe (some 20 years after their first attempt at a four-wheeled car).

Development of the rotary engine began in earnest in the early 60's by a team of engineers affectionately called Mazda's 47 Samurais after licensing rights were secured from NSU & Felix Wankel. The first rotary engine car, the two-seat Cosmo Sport went on sale in 1967 in Japan. It did not garner enough support to be practical so Mazda responded with the four-seat R100 (sold in the USA as model years 1971-1972).

Just as Mazda's dream of separating themselves from their (Japanese) competition via the rotary engine was coming to fruition, the gas crisis of the 70's came about. Had it not been for their concurrent investment in the piston engine, Mazda would likely not exist today. No worries, Mazda repositioned the rotary as a sports car engine and relaunched it in the USA in the first generation Rx7 in 1978.

Fast forward to the late 1980's and Mazda again changed the game with the launch of the $14,000 Miata. A sports car for the 'working' man, the nerve! During the 80's a deep partnership was formed between Ford and Mazda. Several small cars from the late eighties until today share components, some Ford and Mazda's were even made in the same factory.

BROTHERS FROM ANOTHER MOTHER	
MAZDA	**FORD**
1988-1992 Mx6	1988-1992 Ford Probe
1990-1994 Protégé	1990-95 Escort GT
1993-1996 Mx6	1993-1997 Probe
1994-2010 B-Series Pickup	1994-2010 Ranger Pickup
2003-2008 Mazda 6	2006-2011 Ford Fusion

MAYBE NOT MARRIED, BUT DEFINITELY DATING	
MAZDA	**FORD**
2004-2009 Mazda3	1998-2005 Focus
2010-201X Mazda3	2005-2011 Focus
2007-201X Mazda2	2008-201X Fiesta

As the rotary engine went into hibernation in the US market (1995) Mazda introduced new piston engine technology, the Miller Cycle, that was pretty neat but never caught on. The Millenia is the only retail model to have that technology stateside.

I'm of the opinion that Mazda went into hibernation from around 1995 through 2004. Most of the models were suspiciously pedestrian and were lost in an ocean of sameness. This is the reason I've decided to cover 2004+ models in this book. The US market got the significantly improved '3' and '6', the Rx8, as well as Mazdaspeed trim levels for the Miata, 3, and 6 in the years from 2004-2007.

As a side note, it's worth pointing out Mazda has always been an underdog of sorts, David versus Goliath. This is part of the reason I find the brand so compelling. When you consider the variety of engine technologies this little company has explored, it's a testament to their ingenuity and tenacity that they are still around. But more on engine technologies in the next section of this chapter.

5.5 Introduction to Mazda Engine Technology

It's not possible, in the small space I have available for this topic, to do justice to the amount of work nor the technological advancements Mazda and its engineers have contributed to automobile power plants. I'm listing the major engine technologies as I see them with a few interesting facts.

Piston

- **4-Cylinder.** Mazda has been building 4-cylinder engines since the early 1960's. As long as I have been driving, I've known them to be very reliable.

- **6-Cylinder.** Mazda first released a V6 engine in 86 with their J-series engine. The next evolution of their V6, the K-series, appeared as the smallest V6 put in a passenger car, a 1.8L version found in the Mx3 model. Mazdas adoption of Ford Duratec technology represents their V6 engine from the K-series until today.

While Mazda piston engines are generally very reliable, I don't see much to differentiate them from the myriad other piston engine manufacturers.

Diesel

Mazda has never released a diesel engine in North America. However, they have been selling them in most other markets for decades.

Rotary

- With its start in the Japanese model Cosmo Sport through the Rx8 model still sold today, the naturally aspirated rotary engine has a long and storied past. Rotary engines were capable of 100HP per liter of displacement long before this was practical in a piston engine. In fact, to this day there are very few naturally aspirated production gasoline piston engines that can beat the 1971 Mazda R100 for horsepower per liter (displacement).

- Commercially available in the USA from 1987-1995. The addition of a turbocharger totally changes the rotary engine. Not known for their torque, naturally aspirated rotaries can be uninspiring below 4000 RPM. Turbo rotaries in racing have produced upwards of 1000 HP from a 1.3L engine, with 500-600 HP being achievable on pump gas in a street car. Enough said.

- Why haven't these awesome engine's caught on; Two things. First, their fuel economy is not great. Second, they are not as reliable as piston engines. However, if properly cared for, they can be reliable engines. Considering how many companies invest in piston technology, it's amazing Mazda has kept the rotary engine alive, basically alone.

Miller-Cycle

Mazda's only supercharged production engine in the USA was a 2.3L K-series engine producing 210 HP. Using Miller Cycle technology, it works by reducing the amount of energy consumed during the compression stroke of a piston engine. It does this via an ingenious method of 'shortening' the compression stroke.

Direct Injected Gasoline Engines

- **DISI Turbo:** Mazda introduced their direct injected gasoline engine stateside in the Mazdaspeed6 and shortly thereafter in the Mazdaspeed3 and Cx7. Producing somewhere around 260 HP and 280 TQ this engine responds very well to modifications, as most turbocharged engines do.

- **SkyActiv:** Though most people might think of SkyActiv as an engine technology, it's really a more holistic ideal to maximize fuel efficiency without compromising performance. From an engine perspective there are a few things that are atypical from your 'normal' gasoline engine. The SkyActiv engines have very high compression (~13-14:1) and are direct injected (very high fuel pressure).

5.6 Mazda in Racing

Mazda has had a long history of racing what they sell. Starting in the late 60s Mazda first competed with the rotary powered Cosmo 110s, R100s, Rx2, and Rx3. Mazda's real success came with the introduction of the Rx7 in 1978. The first race for the Rx7 was the 24 hours of Daytona winning the GTU class and placing 5th overall, an incredible accomplishment.

Starting in 1980 the Mazda Rx7 was GTU manufacturers champion for 7 straight years. In 1982 Mazda won with the Rx7 at the 24hours of

Daytona in the GTU class and proceeded to continue winning for the next 12 years. The Rx7 is the most winning car in IMSA history hitting its 100th victory in Sept 1990.

Mazda also has a long history in Le mans racing. Sigma, a Japanese race team, entered a rotary powered vehicle of their own construction in 1973 and in 1974. Sigma had assistance from a tuning company named Mazdaspeed which was based in Tokyo. Mazdaspeed took over from Sigma and worked on development and racing of rotary powered vehicles to compete at Le mans and competed almost every year further refining their vehicles and engines.

In 1983 Mazdaspeed built a new vehicle, the 717C, for the new Group C junior category. Mazdaspeed won its class at Le mans and finished 12th place overall in the race. Mazda had been gaining control of Mazdaspeed and by 1984 owned 80 percent of the company. In 1985 Mazdaspeed raced three 737s while work was already starting on the 747.

Jump ahead to 1989 Mazda and Mazdaspeed scored an IMSA GTP class win at Le mans. Director of product planning and development, Yasuo Tatsutomi was not happy with winning the "junior class". After further development and refinements to the 787, the 787b chassis was completed and campaigned in 1991, car #55 finished Le mans in 1st place overall. To date Mazda is the only Japanese company to ever win Le mans and remains one of the proudest moments in the company's history.

Mazda 787b

The same time Mazda was racing in Le mans they were also winning the GTO class with the 4 rotor Rx7, despite getting a 100 lb. weight penalty.

Mazda has some piston powdered racing going on as well. In 1986 Mazda produced a 4wd 323 dubbed the GTX in North America for competition in the World Rally Championship (WRC). In 1988 a Belgian named Pascal Gaban, won the title driving a Mazda 323 4wd. Mazda updated the chassis in late 1989 and won again with another Belgian behind the wheel. In late 1991 Mazda beefed up the GTX and produced a limited model called the GTR. The GTR had a more powerful turbocharged engine, upgraded brakes, and gearbox. Even with the upgraded equipment the 323 GTR did not claim anymore championships for Mazda in the WRC.

Mazda then slowed down its direct racing efforts in the mid 90's and started supporting privateers racing Mazda's through their Mazdaspeed Motorsports support program.

Mazda got back into racing with the introduction of the 99-03 generation of the Protégé used in the Speedvision World Challenge with the help of Tripoint Motorsports, a factory backed team. It took several years of competition and a change over to the Mazda 6 chassis before Mazda scored the drivers and manufacturers title in 2007 with Jeff Altenburg.

In 2008 Mazda was back in the victory circle with SpeedSource winning the Daytona 24hour race in a 3 rotor GrandAm GT class Rx8. In 2010 the SpeedSource team took the team and drivers GrandAm championship in the Rx8 and has continued to campaign the Rx8s until the end of the 2013 season.

In 2012 Mazda saw more success, this time in the Touring Car class of the SCCA Pirelli World Challenge with Michael Cooper driving a Mazdaspeed 3. Michael won the drivers title; and almost single handedly brought home the manufacturers title for Mazda.

In the Continental Tire Sports Car challenge Mazdaspeed 3s and Mx5s occupied the top spots for teams competing in the Continental Tire ST class and just barely missed the manufacturer's title. The I-Moto team won the team championship running Mazdaspeed 3s for the season and looks to return in 2013 to try for a repeat.

For 2013 the big news is Mazda and SpeedSource running the brand new Mazda 6 body cars with the SkyActiv-D engines in the Grand-Am GX class. It should be very interesting as well because this will be the first diesel powered vehicle to run on the GrandAm series and Mazda looks to expand on its number of victories at the 24hours of Daytona.

Please note this is in no way a complete history, I chose specific events and notable points in history to keep it brief.

5.7 Get out and Race (legally)

Now that you have upgraded your Mazda what do you do next? The answer is easy if you want to take it out and see how it performs legally. Race! There are several organizations to help you enjoy safely racing your Mazda. The largest group in America is the Sports Car Club of America or SCCA. The SCCA is full of enthusiasts who love to race their cars in sanctioned events like Autocross, Rally Cross and Club Racing. The best part about the first 2 types is that they require nothing except that you to show up in your car with a helmet (they even have loaner helmets).

- **Autocross** is a course setup with cones, usually in a large parking lot, that define the race path. The idea is to go as fast as you can then see who has the best time compared to similar cars competing that day. Most events have 3-5 runs per driver (you can have multiple drivers per car). The runs are as shorts as 30 seconds and can be as long as 2 minutes depending on space available. There are many different groups or classes to compete in; everything from your mom's stock MPV to a race ready Rx8 can run. After just 2-3 events you will notice an improvement in your driving habits and car control. Bonus for Autocross is it's cheap; most events are around $25 for half a day.

- **Rallycross** is an Autocross but on dirt, gravel, mud, or grass. I participated in Rallycross for several years in my old 323GTX and I can tell you it was a blast. The only downside was cleaning my car after spending an entire day racing my brains out in the dirt. Like Autocross, Rallycross is cheap to participate at $25-$35 for ½ a day.

- **Club racing** is an entry into road racing. It allows you to go head to head with other people on closed tracks to see who is the fastest. There is up-front work for club racing as there is safety gear you have to buy for yourself and your car as well as a special license required to race. There are very specific rules you have to follow for the racing class. The good part is most local SCCA branches have a novice driving school at the beginning of every year to aid you in getting a novice permit and learning the rules.

Conclusion

So there you have it. We've looked at the different platforms available to you, what the different parts are, where they go, what they do, what the benefits are, where to get them, and where to go for more info. I've pointed out things to watch out for, explained common terms and done my best to equip you with the basic information to step out in to the modding world without too much stress. We've also discussed what your options are once you've done all of the entry level mods and you still want more, and how to get your Mazda to "the next level" if that's where you want to go. I hope this book has left you feeling informed and educated about entry level mods and you at least have a basic understanding of what mods might fit best with your needs and lifestyle. I think there is only one step left and that is to get to modifying and enjoying your Mazda.

APPENDIX A

MAZDA COMPONENT CALL-OUTS

Now that you are familiar with all the modifications for both suspension and engine components, it will be useful to know where they are on your Mazda. The pages to follow show common 2004+ engines and where several common engine components are located. The first picture is of a front suspension. While each car will look slightly different, most cars will have the components in a similar arrangement and location. After that you will see a turbo motor example that calls out all major components and a few turbo specific components. Following this is a bunch of naturally aspirated motors common in late model Mazdas.

FRONT SUSPENSION COMPONENTS

1. Strut or Shock
2. Spring
3. Brake Caliper
4. Brake Rotor
5. Brake Line
6. ABS Line

2.3 DISI TURBO (MS3, MS6, CX7)

1. MAF
2. Air Filter
3. ECU
4. Shock/Strut Tower
5. By-Pass Valve
6. Top Mount Intercooler

2.0/2.5 SKYACTIV

1. MAF
2. Air Filter
3. ECU
4. Shock/Strut Tower

2.0/2.3/2.5

1 *MAF*
2 *Air Filter*
3 *ECU*
4 *Shock/Strut Tower*

1.5 LITER (MAZDA2)

1. MAF
2. Air Filter
3. ECU
4. Shock/Strut Tower

3.0/3.5 V6

1. MAF
2. Air Filter
3. Shock/Strut Tower

MIATA

1. MAF
2. Air Filter
3. Shock/Strut Tower

APPENDIX B

GLOSSARY

Backpressure: Backpressure is the resistance from the exhaust system that slows exhaust gases exiting the engine. Some backpressure can be helpful on a non-turbocharged engine. Ideally, you want to eliminate all backpressure on a turbocharged engine.

Boost: No, I am not talking about stealing cars, "boost" refers to the amount the intake manifold pressure exceeds the atmospheric pressure. This represents the extra air pressure that is achieved over what would be achieved without the forced induction (turbo). The level of boost may be shown on a pressure gauge, usually in bar, or psi.

BT (Big Turbo): This is exactly as it sounds, a turbo bigger than the stock one. This mod really implies you are gunning for some larger power #s.

Cat: Catalytic Convertor. A vehicle emissions filter that converts toxic byproducts from your exhaust into less toxic substances.

CEL: Check Engine Light. This is that annoying little light on your dash. You can find out what is causing the CEL by plugging in a code reader to get an engine code.

FMIC: Front Mount Intercooler. This refers to an intercooler that is mounted in front of the radiator (behind the bumper).

Fuel Trim: If your fuel trim goes outside the range that has been set by the ECU, you can expect a CEL to pop up on your dash. There are two trims the ECU looks at, STFT (short term fuel trim) and LTFT (long term fuel trim). The STFT is the quick changes your fuel trims are making as the ECU sees them. The LTFT are the fuel trims that are recorded over time that the ECU "learns" and adjusts to, accordingly.

Hard Parts: These are parts that don't generally wear out. Soft parts are things like rotors, brake pads, and clutches.

JDM: Japanese Domestic Market. aka Made in Japan and sold as original equipment on a Japanese car.

Knock: Knock? Who's there? Knock, also known as pre-detonation or ping, is a sure fire way to ruin an engine and is very dangerous to the life of your motor if not addressed. Knock is where a hot spot in the combustion chamber ignites the fuel instead of the spark plug, resulting in a shockwave; which can have a pinging or knocking sound.

MAF: Mass Air Flow sensor, aka airflow meter, VAF, or pressure sensor. A component in your Mazda used to measure the amount of air entering your engine.

Newb: aka newbie or noob. New guy or beginner.

OEM: Original Equipment Manufacturer, also known as stock parts or those supplied by Mazda from the factory.

Oversteer: aka "loose". The best explanation I have heard used to describe oversteer is "the back of the car hits the wall first", which means your tires lose grip and the back of the car will start to slide first. People who "drift" use oversteer to get the car sideways in a controlled slide.

Running Lean: Too much air vs. fuel in the cylinder of your motor. This can result in knock.

Running Rich: Too much fuel vs. air in the cylinder of your motor. This can result in carbon build up from not being able to burn off all of the fuel. When you see a black puff of smoke from your exhaust, you are running rich.

TIP: Turbo Inlet pipe. The TIP is the pipe which allows air to go from the intake box to the inlet of the turbocharger. Upgraded units provide a smoother path for air entering the turbo which results in more power.

TMIC: Top Mount Intercooler. This refers to the intercooler that is mounted on top of the engine (like the Mazdaspeed 3 or the older FC Rx7s).

Understeer: aka "push". Understeer is when the front tires loose traction first (the front of the car hits the wall first).

WOT: Wide Open Throttle. The gas pedal is pushed to the floor!

BIBLIOGRAPHY

ACKNOWLEDGMENTS

Goodyear Tire

Justin Metz provided great content for the mod cards on tires

Mackin Industries

I've purchased several sets of awesome wheels from Mackin over the years. Justin Banner did a great job of explaining what to keep in mind when shopping for new wheels for your car.

Mackin Industries
9921 Jordan Circle
Santa Fe Springs, CA 90670
TEL: 562-946-6820
www.mackin-ind.com

Robert Finkle

Ever a resource on the forums and just a great person, I've learned a lot from Robert and respect his contribution to the Mazda community.

CorkSport Team

I would like to thank the entire CorkSport team. These people were especially helpful to me in putting this book together.

Gwynne Knowles: Marketing
Brydon Foster: Engineering
Dustin Berentsen: Management
Joel Tandberg: Sales Director

CorkSport Mazda Performance
3315 NE 112th Ave B-80
Vancouver, WA 98682 USA
TEL: 360-260-2675
www.corksport.com

INDEX

B

C

D

E

F

H

I

L

M

N

O

R

S

T

W

MOD Card Index List

ABOUT THE AUTHOR

Who is this Derrick Ambrose character? I have been a car enthusiast as long as I can remember. From slot cars and hot wheels to my first car at age 16 there has always been something about cars that has drawn me towards them. Sure, there have been a few distractions along the way but the draw (dare I say addiction) has always brought my focus back to Mazdas. The pivotal car point in my life was getting a ride to school in a (then new) 1988 323GTX which was a demo car from a local Mazda dealership. It didn't matter what the weather was doing I was always able to make it to school in that cool looking 323. When you throw in

several trips to the IMSA races at Portland Intl Raceway watching the GTO and GTU Rx7s and Firehawk challenge 323 GTs racing there wasn't any chance of buying another brand. In 1998 I started Cork-Sport with my life long friend Richard Harris. Our goal was to supply other enthusiasts performance products from Japan with brands like Mazdaspeed and AutoExe. For over 20 years I have been modifying Mazdas including racecar preparation and questionably legal - street driven build ups with custom tunes. I currently answer the phone every day as part of the sales team at CorkSport and daily drive a 1987 Rx7 Turbo. Last year I got my start in road racing by building the CorkSport B-Spec Mazda 2 and have been able to check one more thing off my bucket list!

Derrick Ambrose
Sales & Co-Owner,
CorkSport Mazda Performance